Algorithmic Barriers Falling: P=NP?

Donald E. Knuth
Edgar G. Daylight

A conversation with Donald E. Knuth conducted by Edgar G. Daylight on 18 June 2014 in Paris, France.

Edited by Kurt De Grave.

LONELY SCHOLAR™
SCIENTIFIC BOOKS

Conversations issue 4
First edition
Version 1.0

© 2014 Edgar G. Daylight
Cover design © 2014 Kurt De Grave
Back cover photo © 2014 Edgar G. Daylight

Daylight can be contacted at egdaylight@dijkstrascry.com.

Published by Lonely Scholar bvba
Dr. Van De Perrestraat 127
2440 Geel
Belgium
http://www.lonelyscholar.com

Typeset in LaTeX

D/2014/12.695/1
ISBN 978-94-9138-604-6
ISSN 2034-5976
NUR 980, 686

Contents

Preface   1

Biography   3

1   Paris   4

2   History   9

3   Asymptotics   18

4   Grammars   41

5   Algorithmic Barriers Falling   48

6   Intermezzo   63

7   More about the 1970s   68

8   TeX & Literate Programming   80

9   P = NP   89

Bibliography   101

Index   112

Preface

The official site of ACM Turing Award winners describes Donald E. Knuth as the rare theoretician who writes many lines of code every day. His main life goal from the 1960s onwards is, in a nutshell, to nail the costs of computation down to the last penny.

Knuth's school of computer scientists has no time for impractical asymptotic claims, nor room for Bourbaki's devotion to infinite constructions. In fact, tensions between the abstract and the concrete led Knuth in 1968 to switch from being a math professor to become a computer science professor. By developing his own "concrete mathematics" at Stanford University, he paved the way for answering quantitative questions about the exact cost of running different algorithms on real computers.

My aim here is to place Knuth's analysis of algorithms into a historical perspective. For this reason, I asked him to contrast his views on asymptotic methods, models of computation, pointer-based data structures, and the like with those of his contemporaries, particularly the theorists including Juris Hartmanis, Stephen Cook, Alfred Aho, and Robert Tarjan.

Meta-considerations constitute another theme throughout the discussion: Why should we historians and computer scientists reflect on the past in the first place, and how can we do so effectively?

The result of this effort is the present booklet, which covers several topics including incentives to study history, events that led to algorithmic discoveries, the beginnings of TeX, and Knuth's evolving views on the P versus NP question. Many topics in this booklet are interrelated. No attempt is made to disentangle

1

them. On the contrary, my aspiration is to analyze a large part of computing's past from multiple angles simultaneously. For example, I look at the creation of TEX in other ways than as seen in the documentation of TEX by experts in digital typography. By discussing the making of TEX in the broadest possible terms, other topics can intermittently take center stage such as Knuth's stance on the P versus NP question in a 1985 paper describing an implementation technique of TEX. Another example is Knuth's passion for analyzing past developments in science and technology — a passion which becomes apparent when Knuth explains how the work of two masters in analog typography influenced his research on digital typography.

By analyzing several writings of Knuth from these complementary angles, we also start to get some grip on the incentives for Knuth to *reverse* his opinion with regard to the P versus NP question, and we begin to see multiple reasons why historical awareness is so important to Knuth and to society in general.

The outcome of this booklet, then, is the painting of a vast landscape, linking Knuth's devotion to the algorithmic, the finite, and the programmable machine to his strong aptitude for mathematics and asymptotics in particular. Seeking the right balance between practice and theory — between the finite, real computer and the infinite Turing machine — is a research agenda that many computer scientists share (cf. Mahoney [94, p.133]). Knuth's role in this regard is vividly told here by no one other than Knuth himself.

— Edgar G. Daylight
27 September 2014, Geel, Belgium

Acknowledgment
I thank Dave Walden for proofreading a draft version of this book.

Biography

Donald Ervin Knuth was born on January 10, 1938, in Milwaukee, Wisconsin, USA. He received Bachelor's and Master's degrees in Mathematics from the Case Institute of Technology in 1960 and a Ph.D. degree in Mathematics from the California Institute of Technology in 1963. After working as an Assistant Professor of Mathematics and Associate Professor of Mathematics at the California Institute of Technology, he became Professor of Computer Science at Stanford University in 1968. Between 1960 and 1968, he also served as a consultant to the Burroughs Corporation in Pasadena, California. He has remained on the faculty of Stanford University up till this day, and is currently Professor of The Art of Computer Programming, Emeritus. Knuth has received several awards throughout his career, including the ACM Turing Award in 1974, the Kyoto Prize in 1996, and the BBVA Frontiers of Knowledge Award for Information and Computation Technologies in 2010.

1. Paris

Edgar Daylight had a rendezvous with Donald Knuth in Paris on Wednesday morning, 18 June 2014. They met at the *25th International Conference on Probabilistic, Combinatorial and Asymptotic Methods for the Analysis of Algorithms* (AofA2014), which was held between 16 and 20 June. Don had helped to launch that conference two days earlier by presenting the first Flajolet Lecture, an invited talk honoring the memory of Philippe Flajolet and intended to be a regular feature of this conference in the future. After attending Wednesday's keynote by Manuel Kauers, entitled 'Analysis of summation algorithms,' Don and Edgar walked and talked through Paris, from the conference building to the apartment where the organizers had arranged for his family to stay during the week. The topics discussed during their stroll covered a wide range, from the analysis of algorithms to taxi strikes in Europe. Part of the outdoor discussion served a preparatory role for the follow-up indoor conversation.

Don: In the 1960s somebody asked me "What do you do?" and so I said "Computer Science" and so he said "Oh, Artificial Intelligence?" and I said "Well, not really." Then he said "Numerical Analysis?" and I said "No." He replied: "So you must be a language man." Evidently Computer Science, like Gaul at the time of Julius Cæsar, was divided into three parts; but I didn't actually live in any of those parts! So I started thinking about a name, and eventually came up with "Analysis of Algorithms."

Suppose you've got two methods and you want to prove that one is going to be twice as fast as the other. I found that this sort of study required mathematical techniques that were mostly different from what I had learned as a student. Richard Bellman had introduced new kinds of mathematics in the work that he was doing, and he'd decided to call it "Dynamic Programming." His work had nothing to do with computer programming, but people nevertheless understood it as programming in the sense of operations research; that is, trying to find a good program for a company rather than for a computer. I said to myself: "Oh, Richard Bellman has made up a name for what he does. Well, I'm going to make up a name for what I do." And I decided that I'd like to call my work "Analysis of Algorithms."

I didn't really have a definition, only that title. For the next two or three years, the working definition of Analysis of Algorithms was "whatever I like to do." But I gradually started to refine it. I asked my publishers if I should change the title of my book from *The Art of Computer Programming* to *The Analysis of Algorithms*; they said no, because that would never sell.

I was developing the mathematics that could be used to answer quantitative questions about the exact cost of running different algorithms. Then I was asked to give a talk at the IFIP Congress in 1971. Every four years there was an international IFIP congress where computer scientists from all over the world would get together. The title of my talk was "Mathematical Analysis of Algorithms." In that talk [56] I gave a definition, which I divided into two parts. One of those parts was the analysis of particular algorithms (for example, the analysis of quicksort); the other was the analysis of a whole family of algorithms (for example, the analysis of all possible ways to sort, or of all the ways to write a compiler). The second part is what became Complexity Theory, and that's something of which I've become more an observer than an active contributor. The first kind of algorithmic analysis, the analysis of particular algorithms, requires different methods. The people who are best at such studies tend to belong to a different kind of community, and those are the people who are gathered here in Paris this week.

Here's the mosque of Paris. We're ten minutes from my apartment.

Edgar: Is there a reason why this conference is held in Paris? Is this way of thinking something the French are good in?

Don: It's universal. There are many A-of-A types all over the world. But quite a few of them are French, of course. We had meetings in Dagstuhl during the 90s; Dagstuhl is where I met you last time (cf. [88]). But the people in Dagstuhl weren't excited about making a permanent commitment because there weren't quite enough Germans in attendance [106].

The man who was sitting behind me this morning is from Belgium. Like me, he's getting older now. I met him at that first Dagstuhl meeting. I'm trying to think of his name ... [Guy Louchard]

Edgar: I think most Belgian computer scientists that have made some name come from the French speaking part of Belgium.

Don: He is French speaking. His specialty is relating algorithms to Brownian motion, which is something biologists began to study in the 19th century. It's a fascinating subject, and he found that a lot of things that come up with algorithms can be understood if you connect them to Brownian motion.

Edgar: Have you been on trips to Southeast Asia during your career? Have you seen the state of the art in computing in those countries? I know you have been to Eastern Europe.

Don: There are many countries where I've never been. I mostly stay home to write. I probably turn down two opportunities to travel every week! This conference in Paris is an exception. My co-worker Philippe Flajolet died suddenly three years ago [103], so the community decided to have a series of lectures in his honor here in Paris. I gave the first one on Monday. I'm happy to see that the subject of analysis of algorithms is thriving. There are hundreds of young people solving problems today that I was never able to solve myself.

I went to China in 1977 ... also to Japan once and Singapore twice. I've never been to Australia or New Zealand. I was in South America once. I've been in Armenia and Uzbekistan. I haven't been to Africa. (I did once see Egypt from the other side of the border when I was in Israel, and I once glimpsed Morocco from an ocean liner.)

Some people in China asked me to join the editorial board of a journal. This was long, long ago (1988) when there was almost no computer science in that country. It's called the *Journal of Computer Science and Technology*, and the journal has grown significantly in quality ...

Edgar: Can you tell me about Vaughan Pratt? Was he your PhD student?

Don: He's Australian. He studied linguistics as an undergrad there, then came to Berkeley for graduate studies in computer science. I met him one day and persuaded him to switch to Stanford, because we had so many common interests. He certainly was the most versatile student I've ever had. He finished his PhD in record time. Now he's my colleague, with an office right next door to my own. These days he tries to teach *me* about things.

Don and Edgar enter Don's apartment.

Don: Now we can walk up two staircases, to the second floor. That's 19 plus 17 steps.

Edgar meets Don's wife, Jill, and one of their grandchildren, Kadin. Jill talks to Edgar about Paris.

Jill: We were trying to ride buses rather than the metro because you can see more. But we did get caught yesterday in the Uber demonstrations. The traffic had been diverted and buses weren't running in the place where we hoped to get a bus.

Edgar: There was a strike?

Jill: Taxis are on strike all over Europe because of the Uber threat. Do you know this whole story?

Edgar: No.

Jill: Uber is a phone app that allows you to get rides on cars from private people. It started in San Francisco. Because it is unregulated, the drivers don't pay the same taxes and they don't need the same licenses as taxi drivers. Therefore the rides are a lot cheaper. You look at your phone and you say I'm here and I want to go there and here's a car that's willing to take me for a small fee. I guess it's very effective. We have a friend who goes to Washington D.C., and he says you can't get along without it. Taxis are impossible to find but you get around that way.

Edgar: I'm really out of touch. I didn't know about this.

Jill: At any rate, the taxi drivers all over Europe are naturally very unhappy about it. So it's a case of new technology taking over the old ways.

Edgar: Yes, it's coming anyway.

Jill: Right. Taxi drivers protest by going on strike. Also, a whole lot of taxi drivers congregate at a certain place and demonstrate or cause a traffic jam. They had probably hundreds of police vans and police in riot gear. We saw a small fire, lots of taxis, and no buses.

Edgar: It's strange I didn't see any of this in the news.

Jill: It's been in London, Barcelona, Berlin, and Paris. So we've had to walk a lot to get to where we intended to go.

2. History

Don and Edgar start discussing in the apartment's living room.

Don: A month or two ago I gave a talk at Stanford that was called 'Let's Not Dumb Down the History of Computer Science' [86]. As you know, I'd discussed my concerns with you already [88, Ch.6], and this time I went public with my comments about why history of science is so important.

Edgar: There's a Special Interest Group on Computers, Information, and Society (SIGCIS) with an email list for historians of computing.[1] Several historians have been discussing your talk, even before the video recording of your presentation was made available online.[2] So it had an impact.

Don: The responses might be positive or negative, but anyway I stated my point of view. I'm glad because the previous time, at the University of Greenwich, my comments were lost due to a faulty recording mechanism.

Edgar: The reactions to your talk certainly aren't negative. There's a bit of everything. The community of historians is small though. I believe there were three weeks of discussion on the mailing list and then it calmed down again. That said, I am aware of several discussions still going on privately.

Don: The ideal way I think to present history is to zoom in on some technical topics and explain them to people who

---

[1]See *www.sigcis.org*
[2]See *kailathlecture.stanford.edu/2014KailathLecture.html*

aren't specialists, and also to zoom out and paint the broad picture. But not to skirt from scientific stories, from the technical details. Every piece of science can be explained to nonspecialists, if you take the time to do it.

Edgar: Then there's the question of *how* to present a piece of science. I notice that you intentionally include mistakes in your writings. For example, in Volume 1 of *The Art of Computer Programming*, you wrote that:

> The alert reader will observe that we have made a serious mistake in operations (a), (b), and (c). What is it? Answer: We have forgotten to consider the possibility of an *empty* list. [75, p.274]

So you "overlooked" the *empty* list in your book for didactic purposes?

Don: The art of doing science is the art of learning how to recover from mistakes! You admit them and you understand how to move forward. Historical accounts make it clear that the smartest people in the world have continually made mistakes. This tells me not that they were stupid but that the ideas were not trivial. Even great minds got them wrong at first. So we can learn about the subtlety of concepts by seeing how many people got them wrong. Once we are more familiar with a subject, we can condition ourselves to avoid those errors. It takes time to reach this point. So every once in a while it is good to realize that we're working at a certain level of subtlety. That happens with all aspects of science, and certainly with algorithms and programming languages. That's also why I pointed out in my talk that I like to read source materials from past years.

My little book *Surreal Numbers* [62] demonstrates repeatedly how to recover from mistakes. It features two characters, Alice and Bob, who work out a theory, and it's a dialog between the two of them. Alice has more abstract abilities to see large-scale issues; Bob has more brute-force capabilities to carry through. They work together but also they make errors and recover from them. The mistakes that they make are errors that I myself made as I was working out the theory. So I faithfully recorded my blunders and ascribed them to the characters in the story.

I taught a class about combinatorial mathematics during my first year at Stanford (1970), and my style of teaching (learned from George Pólya) was to have conversations with the students instead of telling them what to write in their notebooks. During the first or second session we discussed a particular combinatorial pattern, and I asked the students to guess how many ways there were to get from *A* to *B*. They gave wrong answers; but then I had an inspiration: I said, "Well, we can check a few cases and see that this is not the right answer. But is there *another* problem for which our answer *would* be correct?" In other words, we could make use of our error because it led to another question about which we could learn more. I certainly think errors are incredibly important in developing science.

Edgar: Peter Naur has discussed this matter as well. He says he was staying at your home and you arranged a meeting for him with Pólya [18, p.45].

Don: It could be. Pólya became a friend of ours at Stanford. Although he was in his nineties, he would chase our children up and down the stairs! He was one of the greatest mathematicians ever.

Edgar: You have written that:

> To do well in mathematics, you should learn methods and not results. And you should learn how the methods were invented. [2, p.202]

Don: I'm sure Pólya has expressed the same sentiment. It explains why I don't like the way scientific papers are presented now. They have an abstract, and the abstract mentions only the results. It doesn't tell what methods are used to get these results. Sometimes the abstract does say, "By the way, we proved this by using a method that might be of interest in itself." But that's very unusual; the abstract will usually only announce the bottom line. People rate a paper by this bottom line instead of by its methods. What you really learn is the method.

The problem that's solved in a typical paper is usually not the problem you want to solve yourself; it's close to your problem, but not quite the same. So you have to twiddle the method a little bit in order to get the answer you seek. If you only have a black box that says "Given A, B, and C, then such and such follows," and you close the box and hide the steps by which the result is derived, you can't solve any problem that doesn't exactly match A, B, and C. The method is really key.

Edgar: This brings me to your work on TEX in a rather strange way. You studied the work of two masters in typography: Goudy and Zapf. In 1996 you said that

> engineers often make the mistake of not looking at the traditions of the past [. . .] [I] spent May and June reading the works of Goudy and Zapf and everything I could find, back through history. [. . .] It was fascinating, it was wonderful, but I also wanted to make sure that I could capture as well as possible the knowledge of past generations in computer form. [15, p.359]

So one motivation to study the past is to further current-day technology?

Don: Yes. Another is that it's a damn good story. A historical account has got some drama to it.

Let's consider science in general. In my talk I forgot to mention that I greatly enjoyed reading Isaac Asimov's two-volume introduction to the history of science [5] when I was young. It was completely different from any other books of its kind: He not only explained all of the science, he also talked about the people who were involved; he brought everything into a human context. Other introductions to science would either neglect history entirely, or they might perhaps say "In 1955 so-and-so discovered such-and-such about electrons"; but they would give no indication that some human being actually grappled with these problems while working patiently in a laboratory, having numerous failures and so on. Asimov was a master at weaving together the combination of internal and external history that I like.

Edgar: In your talk you said there are six reasons to study the history of a field. We've just discussed the first and second reasons:

1. Understanding the process of discovery.
2. Understanding the process of failure.
3. Celebrating the contributions of many cultures.
4. Story telling: the best way to explain.
5. Learning how to cope with life.
6. Become more familiar with the world.

Don: The first reason is the most important, the second reason is the second most important, and so on.

Edgar: Could you elaborate on the third reason to study history?

Don: I love the fact that we are part of a global community. People have different ways of looking at life. Somehow we come together at the level of loving the science and the patterns that we see, even though we don't share languages, we don't share many traditions, and so on. This is not only true for computer scientists, and for people with many years of formal education; basket weaving, for instance, is a world-wide enterprise. Basket weavers have found beautiful ways to combine different colors of hemp to make pixel patterns that stimulate both halves of our brain — the patterns have a mathematical point to them in some ways. Having so many different contributors from so many cultures — Vive la différence. It's a source of pleasure, a reason to celebrate.

Edgar: The fourth point in the list is story telling.

Don: One of the most remarkable ways in which human beings differ from animals is the fact that we tell stories, and we learn from the past through those stories. For centuries we've been conditioned to pass information from one to another in the form of stories, even if they are just fables. There are Grimm's fairy tales or maybe, say, the story of how the steam engine was invented. All of these have some kind of a plot. That makes it easier for our brains to store the information and to retrieve it. People who teach tricks for memorizing random tables of facts say that the best strategy is to make the data into a story. You identify each word with a room in the house or something, then you tell a story about walking around the house. People have memorized 200 digits of pi by making up a story about them.

Edgar: Then we have the last two points in your list: learning how to cope with life and becoming more familiar with the world.

Don: That's external history.

Edgar: Since the first four points are about internal history, I'd like to discuss the possibility of organizing courses on the history of computing that are funded by computer science departments. These departments have a lot of money, compared to departments in the humanities where history is normally centered. Do you expect computer science curricula to contain such courses in the near future? I suspect that it already happens more in other fields, such as physics.

Don: When I was an undergraduate at Case I co-edited a magazine that was a student publication, called *Engineering and Science Review* (cf. [46]), which came out four times each year and had student-written articles about scientific topics. So it was kind of a training ground for technical writing, and we included historical topics. I took turns with a guy called Charles Weiner on editing that magazine. He later became the main historian for the *American Institute of Physics* (AIP), the principal organization for U.S. physicists. He would visit the widows of famous physicists and acquire their archives on behalf of the society. He wrote a lot of things about the history of physics, so he was full time supported by the AIP. I believe that happened within five or ten years of our graduation in 1960. We chanced to run into each other some ten years later and compared notes. I don't know what has happened to him since then. I only know that the physicists did provide funding for history at one time, although they didn't support universities directly.

But I'm not a source of any wisdom whatsoever about economic things. I've never been able to understand who pays for computers or why they do it. My understanding of university budgets is even less.

The other comment I wanted to make is that writing papers about history is more difficult than writing papers about research. Probably nobody knows how hard such a task is unless they've actually done it. I wrote four papers that were principally devoted to aspects of computer science history [49, 54, 67, 91]. Each of them was much more of an effort than the others on my list of publications, because there were so many more things to track down and get right. In an ordinary paper, the research ideas have one or two logical threads and that's all you need. They don't involve a wide variety of different side threads that must be tied in. Thus it's understandably rather rare that a good historical paper is actually written, no matter by whom. Therefore it is very admirable when it is done.

Throughout my teaching career, I was hoping to come across a student who would be a historian of computer science, someone whom I could sort of help groom when he or she was young. Instead of training that person to be a professor of computer science or a researcher, I wanted to help mold a key historian. During all those years I had just one lead: A young man applied for graduate study in computer science at Stanford, intending to focus on history. I met him and encouraged him, but he went to Princeton. So personally I've failed in that respect.

I'm motivated by good examples, and one of the best that comes to mind immediately is Jeffrey Shallit. He wrote an article about the history of how people have analyzed Euclid's algorithm [104], showing how early researchers had discovered its ties to Fibonacci numbers. The historical notes in Shallit's books are also outstanding. And that reminds me also of Joachim von zur Gathen, who corresponded with me several times during the 90s about source materials he found related to algorithms for polynomial factorizations. The historical notes in his book with Jürgen Gerhard [33] are especially exemplary. (See also [32].) These are people who were writing textbooks about the subject, and realized that there were important chapters of history waiting to be written. They caught the craving that I have, an addiction to reading source materials. But college courses entirely devoted to historical topics are not the kind of thing that Stanford has space for in our curriculum at the moment, except as supplementary material in existing courses or a special topics course to be offered once every few years. The same situation prevails in math departments, although I guess it's not extremely uncommon for them to offer a semester course in history of mathematics.

Edgar: I know many computer scientists who are interested in historical accounts, but are there many people who conduct research the same way you do?

Don: Alas, I was not able to convey the thrills of digging deeply into source materials to any of my own students.

Edgar: In the 1960s you worked closely together with Robert Floyd. Did he have a totally different way, a complementary approach, to tackling a technical problem?

Don: Oh yes. Bob was self taught in everything. He read widely but he mostly would work things out on his own without investigating how anybody else had solved the problem. He had broad interests. But Edsger Dijkstra's interests were even broader: Whatever topic came up in whatever aspect of life, Dijkstra knew a lot more about it than I did.

Edgar: Did Dijkstra go back to the sources?

Don: This I don't know. But his knowledge covered many fields.

I have an extra motivation when I write *The Art of Computer Programming* (TAOCP). Whenever I discuss an idea in TAOCP that I'm pretty sure is going to be interesting 50 years from now, I think it is worthwhile to track down where the idea first appeared in human civilization. That leads me to fascinating search problems. There's nothing I like better than going into a great library with such a problem to solve and then to follow my nose. Now with the Internet we have a great library at our fingertips. Everybody can do this. Every important technique leads to an interesting thing to discover: When did people first have a hint about it?

Yet I have to admit that in all my years of teaching I was not able to transmit this particular research methodology to my students. There might be something wrong with it.

Edgar: That's a fascinating aspect in its own right.

3. Asymptotics

Edgar shows some books to Don who browses through them.

Edgar: Here's Thomas's book *Calculus and Analytic Geometry*. It's the third edition [42]. I was not able to find an earlier edition.

Don: It's from 1961. I was a student in 1956; I probably had the second edition. Where are the supplementary problems? That was the main thing for me. No, you see, my book had a whole section called 'Supplementary Problems.'[1] There were questions at the end of each chapter but then the supplementary problems at the back — I think there were more supplementary problems than problems. I was so frightened that I'd fail the class that I decided to work on the supplementary problems, which weren't assigned by the teacher. I worked very hard at the beginning of the course because I was doing those extra problems. Fortunately, after two or three months of this, I could work the ordinary ones much faster, because I'd become familiar with lots of problems. And with problem-solving skills under my belt, I could coast from that point on.

Thomas's book was one of the two or three that made Addison–Wesley a leading publisher of college texts. 'Addison' and 'Wesley' were the middle names of the two founders, who were printers. They knew some great teachers from MIT and they said, "We would like to print

---

[1]Note by Knuth after returning home: Yes, my copy says "Second edition, fourth printing, February 1956." The supplementary problems appeared on pages 725–792, and their answers appeared on pages 793–814.

your textbook for you"; and they set up a type shop. Their firm was the only textbook publisher that had its own in-house compositors to set mathematics into type. All of the other companies would farm such work out to independent vendors; but right next door to Addison–Wesley's editorial offices was the building that housed Wolf Composition. Hans Wolf was an Addison–Wesley employee. So they did their own typesetting, and they commissioned a designer named Herb Caswell to design their books. Caswell was the designer of this book by Thomas. Hans Wolf trained people to do the typesetting. So Addison–Wesley was the best in the business for quality; I loved their textbooks because they were well produced. When a representative from Addison–Wesley came to me and proposed that I should write a book for them I was immediately intrigued because I had admired the consistent quality. And it was no coincidence that the algorithm-drawn fonts used in TAOCP today were directly inspired by the Monotype fonts used in Thomas's text.

Edgar: I brought some other books and papers from the 1950s to show you for comments.

Don: A. A. Markov's *The Theory of Algorithms* [95] — I didn't read that book until some time in the 1960s. This Markov is the second generation of Markovs, the son of the famous Andrei Markov whose name is enshrined in "Markov processes." The elder Markov not only wrote about probability theory at the beginning of the 20th century, he wrote about continued fractions at the end of the 19th.

H. B. Demuth's *Electronic Data Sorting* [20] — I think I only learned about it in the 1970s, when I was trying to track down the author's middle name. This was his PhD thesis, which was one of the very first works to study questions about the running time of algorithms in general and of sorting algorithms in particular. There was a survey paper by a man named Friend in the ACM journal in the same year [28], which also did some pioneering analyses of sorting.

Burks & Wang's 'The Logic of Automata — Part I' [8]. That one is new to me. I did study Wang's later paper in *Scientific American* [112], which was the inspiration for TAOCP Section 2.3.4.3.

Edgar: I put that title here because we're going to talk about automata theory and Stephen Cook's work.

What about de Bruijn's 1958 book *Asymptotic Methods in Analysis* [7]?

Don: Ah, a wonderful work. I learned about it as a math Caltecher in the middle 1960s. De Bruijn was a good friend of my Dutch professor at Caltech, Wim Luxemburg. Luxemburg, my analysis teacher, was noteworthy because his lectures on analysis were so beautiful — the essence of perfection. I admired everything about them. On the other hand, their elegance convinced me not to work in analysis, because I couldn't imagine ever being able to approach that level of sophistication.

I met de Bruijn when he came to Caltech for a year, in 1965 or so. I was working on the first draft of TAOCP, and trying to analyze every important algorithm that was discussed in the text. I had absolutely no idea in 1965 that 'Dick' and I would become such close friends in later years. I had been working on a problem about algorithms for searching that are based on so-called 'trie' data structures, and I'd found very strange patterns in the behavior of this algorithm. I'd begun with empirical studies, working with tries of different sizes, just to see how long a typical search would take. A trie has two parameters, the number of elements and the degree of branching, and the behavior was quite peculiar when I changed the amount of branching. I worked out the mathematics underlying it as well as I could, but many of the related problems were beyond my reach.

So I asked Luxemburg and he said "Oh, de Bruijn is good at this. Ask him." And de Bruijn taught me some of the basic techniques of complex analysis that were in his book on asymptotic methods. He had discovered beautiful ways to study the behavior of functions as the parameters grow large or small, and he presented them in this unique book, which kind of revolutionized the subject. He had come across these ideas during the Second World War when he had nothing else to do; after studying the works of some other great mathematicians he developed his own nice way of looking at the methods and bringing them into a broad general framework that was widely applicable. So he could tell me brand new ways of looking at problems that were fundamental to the analysis of algorithms, and my eyes were opened. Our friendship blossomed.

Edgar: You were already applying similar techniques in connection with computer programs? [Don: Yes.] De Bruijn's book was not about computer programs.

Don: That's right. But in order to study the computer methods, I wrote down formulas, usually recursive formulas. De Bruijn's methods of asymptotic analysis would apply to these recursive formulas. Say I wanted to know how many ways there are to write an integer n as a sum of powers of 2. For example, take the integer 7. That number is equal to $1+1+1+1+1+1+1$ or $1+1+1+1+1+2$ or $1+1+1+2+2$ or $1+1+1+4$ or $1+2+2+2$ or $1+2+4$; and the number of different ways might tell me the running time of a particular computer program that involved the number n. So I'd count all those ways. For any integer n I needed to know how many ways there are to write n as a sum of powers of 2, say $s(n)$. We can compute $s(n)$ using recursive formulas: $s(0) = 1, s(2n+1) = s(2n) = s(2n-1) + s(n)$. And de Bruijn's methods tell us how to use those formulas to predict the approximate value of $s(n)$ when n is large. The graph of $s(n)$ is interesting, it has "dips" when n passes a power of 2, and we need to understand the behavior of this periodic dipping factor. De Bruijn explained to me how to study this problem by actually working with complex numbers instead of real numbers. There are beautiful theories I had never heard about, one of which is called the Mellin transform. I could transform the problem about $s(n)$ into a problem about a different function that involves complex numbers, and that function wasn't hard to study. Furthermore $s(n)$ turns out to be related to the running time of interesting algorithms.

Edgar: How much was de Bruijn himself involved with computers?

Don: Unknown to me at the time, he had been contributing notes about computing to Dutch educational journals. I later saw papers that he had written in which he gave machine programs to enumerate polyomino tilings and things like that. I could guess enough of the Dutch words and I could read his computer programs, so I knew that he was a programmer in the 1950s. He also made significant contributions to other topics, of which I was unaware until I had known him for almost ten years, because I was talking to him about complex analysis and not about programming. Those aspects of his work were revealed to me much later.

De Bruijn came back to Caltech when he was working on his AUTOMATH project, which was a new twist on mathematical logic, a way of formulating proofs in mathematics. I worked a little bit with him on that. I tried to get him to change AUTOMATH, basically to extend it in a way that we now call object oriented. You can regard the atomic elements of mathematics as objects that have types; for example, if something is of type *integer* you can prove a theorem that is applicable to integers. Another type was *prime number*. The way that he had things set up, if something was a prime number you couldn't use your theorems about integers directly; you had to redevelop them by reproving everything that you'd done for integers. So I encouraged him to make the system more object oriented, exploiting the fact that primes are a subclass of the integers and allowing results about integers to be carried over to primes. But he wanted to keep AUTOMATH simple because it was supposed to be for checking proofs; he preferred to avoid any subtle linguistic details. I was more interested in efficiency of writing the proofs, but he was more interested in the reliability of checking them.

Edgar: I'm particularly interested in your developing notion of asymptotic complexity. You've come back to it throughout your career. Recently, in 2014, you replied to a question posed by Robert Tarjan as follows:

> Most of today's literature is devoted to algorithms that are asymptotically great, but they are helpful only when n exceeds the size of the universe.
>
> [...] I'm emphatically not against pure research, which significantly sharpens our abilities to deal with practical problems and which is interesting in its own right. So I sometimes play asymptotic games. But I sure wouldn't mind seeing a lot more algorithms that I could also use. [110]

Don: A large literature has grown up where people show how to speed up the solution to some problem by saving a factor of, say, $\sqrt{\lg \lg n}$ in the asymptotic running time. Here n is the size of the problem, and the authors consider the rate of growth as n approaches infinity. Now, $\lg \lg n$ is never bigger than 6 in practice,[2] because $\lg \lg n = 6$ when $n = 2^{2^6} = 2^{64}$; that's the number you get when you start with zero in a 64-bit register and add 1 to it as fast as you can until the register overflows. If n is any number that we can represent in one long computer word, then $\lg \lg n < 6$. Now, while saving a factor of the square root of 6 the authors of this typical paper are also increasing the constant in front of $\sqrt{\lg \lg n}$; their method might win asymptotically, but only when $\sqrt{\lg \lg n}$ exceeds 100. But then $\lg \lg n$ exceeds 10000, and n exceeds $2^{2^{10000}}$;

---

[2]In Knuth's books 'lg' denotes the logarithm to base 2, 'ln' to base e, and 'log' to an unspecified base.

this number greatly exceeds the size of the universe, which contains nowhere near even 2^{500} protons.

Theoretical computer scientists wonder whether certain methods are absolutely the fastest possible as n goes to infinity. That's a natural, clean, mathematical problem to ask, and it provides a scoring system in which people can compete with each other. If you can say your method is $\sqrt{\log \log n}$ times faster than previously published methods, you know that your algorithm is new, and you can publish it. But programmers in the real world are usually better off if they stick to simple methods that are asymptotically inferior.

An asymptotic improvement is intellectually and academically satisfying. From the standpoint of TAOCP, however, the value is basically nil — except to show that one couldn't prove that an asymptotic improvement is impossible. So those methods are publishable and have their appeal, but I really have no space for them in TAOCP unless the particular problem domain is especially fundamental.

Edgar: You do use asymptotic methods in your own work and you used them in your first draft of TAOCP in the 1960s, right?

Don: The main difference between the way I tend to use asymptotics and the way most other authors do it is that I use asymptotics only after the leading term. I'll say that something is $2n + O(\sqrt{n})$, instead of simply saying that it is $O(n)$. I'm giving a sharp asymptotic analysis and I'm bounding the error, but I'm not bounding the main term.

When people give asymptotic estimates they use O and o notation, 'big-Oh' and 'little-oh'. Big-Oh notation means that the true value of a quantity is not bigger than a constant multiple of what you're expressing, for some unspecified positive constant. But little-oh notation goes further; it says that the constant can be arbitrarily small, when n is large enough. For example, if I assert that something is $O(n)$, I'm telling you that is less than or equal to Cn but I'm not telling you what C is. And if I assert that something is $o(n)$, I'm telling you that it is less than $.0001n$ when n is large enough, and that it's less than $.0000001n$ when n is larger yet, and so on; any quantity that's $o(n)$ must grow more slowly than n. For instance, $\sqrt{n}$ is $o(n)$, and $n^{.9999}$ is $o(n)$. Saying that a quantity is $o(1)$ is equivalent to saying that it approaches 0.

When I started TAOCP and got to the point that I wanted to define O notation, I decided not to complicate the picture by also defining o notation. In none of the volumes of TAOCP do I ever use o notation. (My book *Concrete Mathematics* [36, 37] does talk about o, however.) This decision meant that I had to learn more mathematics.

You see, I never allow myself to say that a quantity is $o(n)$, so I have to say instead that it is $O(\sqrt{n})$ or something similar; that is, I have to give a sharper estimate for the error. I can't just say that some quantity grows slower than n, I have to say that it doesn't grow faster than some explicit function of n. This choice forced me early on, all the way through, to learn better mathematics, because I was almost never able to take a result out of the literature and put it right into my book. Extra effort was needed, just as when I chose to do supplementary problems in my calculus class. But I found works by people like de Bruijn who had developed techniques for sharper estimates; so I gradually become familiar with O and learned how to do without o. As a result, after a couple of years, I could do a lot of other problems that would have been beyond me if I hadn't set higher goals in the first place.

Edgar: You chose not to use O notation for the leading term. But why did you avoid the o notation?

Don: First it was just to save space in my book. But, later on, I was glad that I had learned these other techniques. I once wrote a little essay entitled 'Teach Calculus With Big

O' [76] based on what I've learned. If I had another life to live, I'd love to rewrite the standard textbooks on calculus in order to show students how beautifully simple it is to use O notation in the middle of formulas. I learned such ideas mostly from mathematicians who were working in number theory. The great news is that O fits neatly into ordinary algebra, and I wish somebody had taught me early on that it's easy to manipulate such formulas. Formulas with O tell us that quantities aren't too big, without bothering to reveal exactly how big they are.

Edgar: Did you, at some earlier stage in your life, use O notation for the leading term?

Don: That's an excellent question. In fact, I can pinpoint fairly well the exact moment when I noticed that something was fundamentally wrong in the existing literature. I was reading a book by Ken Iverson, probably in 1963, . . .

Edgar: Do you mean this one? [He shows Iverson's book *A Programming Language* [41].]

Don: Wow — talk about serendipity! How on earth did you have the foresight to bring that very book with you?! Yes, here it is on page 234, Table 6.36, a table of numbers that changed my life. When I saw that table, I was struck by the realization that it was a bad way to understand good mathematics, I mean a misleading way, although the numbers are technically correct.

Let me try to explain. This part of Iverson's book analyzes the popular bubble sort algorithm, which involves a quantity that he called $z(n)$. We need to know the asymptotic value of $z(n)$ if we want to understand the average number of passes over the data that bubble sort makes before it decides to stop. There's a complicated formula for $z(n)$, but no simple way to express it in terms of more familiar functions.

Some people had managed to prove that $z(n)$ is asymptotically equal to $\sqrt{\pi n/2}$, which means that $z(n)$ divided by $\sqrt{\pi n/2}$ approaches 1 as n goes to infinity. Iverson's table shows the actual value of that ratio, which he called $c(n)$, when n is 1, 5, 10, ..., 50; for example, it says that $c(10) = .877$ and $c(50) = .934$. But I didn't find those numbers very convincing, because convergence to the limit looks pretty slow. Instead of considering what happens when $z(n)$ is *divided* by $\sqrt{\pi n/2}$, I decided to look rather at the difference, $z(n)$ *minus* $\sqrt{\pi n/2}$; and voilá! If we set $d(n) = z(n) - \sqrt{\pi n/2}$ we find that $d(10) = -.484$ and $d(50) = -.585$; indeed, one can prove that $d(n)$ approaches a *constant* value. Thus the right way to understand the asymptotics of $z(n)$ is to know that $z(n) = \sqrt{\pi n/2} + O(1)$. This is a much stronger statement than the previous claim about the ratio, which tells us only that $z(n) = \sqrt{\pi n/2} + o(\sqrt{n})$. Iverson could have presented a more meaningful and more convincing asymptotic analysis if he had tabulated $d(n)$ instead of $c(n)$. I think I stopped using O in the leading term the day after I saw Iverson's Table 6.36.

Eventually this quantity $z(n)$ found its way into Section 1.2.11.3 of TAOCP under the name $P(n)$, where I proved (using de Bruijn's methods) that the asymptotic form of $P(n)$ begins with $\sqrt{\pi n/2} - 2/3$ and continues with terms of order $1/\sqrt{n}, 1/n, 1/n^{3/2}$, etc. (See [75, p.120].)

I had already gotten hooked on the analysis of algorithms about a year earlier. In January of 1962 I was asked to write TAOCP; but I was still a graduate student, I hadn't finished my thesis. I spent the summer of 1962 writing a compiler for FORTRAN, and I got to the part of that project where my compiler was supposed to search for identifiers in a hash table. I said to myself, "Hmm, I've always wondered why hashing works"; so I took a day off from compiler writing and looked more closely at the hashing algorithm. More specifically, I looked at a certain kind of hashing called linear probing, because I'd used linear probing at Case when writing assemblers. I tried to figure why that method always worked so nicely — at least it seemed to work well until the hash table started to get full.

Fortunately, I was lucky and thought of a way to solve that problem, using some mathematical functions I'd never seen before. I typed up some notes on what I'd done; this was the first progress that I had actually made towards writing TAOCP. (A few years ago I sent a copy of those notes to Flajolet, who scanned them and put them on the web.[3] They are dated 27 July 27 1963.) I had heard that some students of professor Feller at Princeton had been working on linear probing, without success. As I said, I was working seven-day weeks, writing a compiler, but I took a day off to play with the hashing problem and happily was able to crack it. Then it occurred to me: "I've heard about Queuing Theory, which is devoted to the analysis of a certain limited class of algorithms. There are lots of varieties of queues. But if we go beyond queues to things like hashing and all the other basic computer methods, each of them must be associated with problems that would be just as fun to solve as the hashing problem I have just licked." That moment was like an epiphany, when I realized that the pursuit of algorithmic analysis would make a good basis for my life's work.

In summary, then, I was learning how to analyze hashing, and I learned about O notation by reading number theory papers. I gradually learned how to manipulate O notation and to become comfortable with operations like the computation of $n + 1 + O(1/\sqrt{n})$ raised to the power of $2 + O(1/n)$. I learned how to do algebraic operations with O as I had learned to do other algebraic operations in high school. This was all happening in 1963 and 1964, when I was starting to write TAOCP. The analysis of hashing went into Chapter 6, in Volume 3, later on.

Edgar: When did people like Tarjan start to use asymptotic methods in their own work?

_____

[3]See *algo.inria.fr/AofA/Research/11-97.html*

Don: Tarjan began graduate study at Stanford about 1970, and was attracted to complexity theory rather than to the precise analysis of particular algorithms. Like other workers in complexity theory, he liked asymptotic results with O in the leading term, mostly because such results are independent of the machine on which an algorithm is implemented.

Consider Hoare's quicksort algorithm, for example [40]. Complexity theorists generally are content to know that its running time is $O(n \log n)$. By using O notation in the leading (and only) term, they can throw all other details under the rug, including the type of machine they have. By contrast, I like to know that quicksort makes $2n \ln n - \frac{8}{3}n + O(\log n)$ comparisons, on the average, and $\frac{1}{3}n \ln n - \frac{4}{9}n + O(\log n)$ exchanges, and $\frac{2}{3}n + O(1)$ stack adjustments, when sorting n random numbers. Instead of lumping everything together and looking at the total, I prefer to rate each individual step, also knowing which steps access memory in which ways, so that I can have a sharper understanding of what's going on.

Edgar: You wanted the mathematics to match the real experience you had with your program. [Don: Yes.] Figure 3.1 contains the rest of your reply to Tarjan. As a mathematician it is tempting to abstract finite data sets and go to infinity.

Don: Mathematicians like to see everything from one unified picture. As a computer scientist, part of me doesn't mind going into cases. A computer scientist is comfortable with Step 1, Step 2, and Step 3 being different. Our talents for dealing with nonuniformity also tend to make us miss unifying results when they exist; but many problems don't have a unifying result and then we can make headway on them while mathematicians throw up their hands.

Edgar: How far do you go in making case distinctions? You mentioned during our stroll through Paris that you prefer to analyze a particular algorithm while complexity theorists typically analyze a family of algorithms.

For instance, I've been reading about algorithms that decide whether or not a given graph G belongs to a certain class. Is G, say, chordal? You and others discovered some great algorithms for the chordality and minimum fillin problems [...] But I've been surprised to discover that very few of these newer algorithms have actually been implemented. *They exist only on paper*, and often with details only sketched.

Two years ago I needed an algorithm to decide whether G is a so-called comparability graph, and was disappointed by what had been published. I believe that all of the supposedly "most efficient" algorithms for that problem are *too complicated to be trustworthy*, even if I had a year to implement one of them.

Thus I think the present state of research in algorithm design *misunderstands the true nature of efficiency.* [...]

Another issue, when we come down to earth, is the efficiency of algorithms on real computers. As part of the Stanford GraphBase project I implemented four algorithms to compute minimum spanning trees of graphs, one of which was the very pretty method that you developed with Cheriton and Karp. Although I was expecting your method to be the winner, because it examines much of the data only half as often as the others, it actually came out two to three times worse than Kruskal's venerable method. Part of the reason was poor cache interaction, but the main cause was *a large constant factor hidden by O notation*

Figure 3.1: A large part of Knuth's response to a question posed by Tarjan in 2014 [110, Daylight's emphasis].

Don: Right, but TAOCP also discusses the complexity of sorting. Suppose you have any sorting algorithm that uses comparisons; then people have known since 1950 that you have to make $n \lg n + O(n)$ comparisons, regardless of your method. (In this formula 'lg' is again short for 'logarithm to base 2'.) To prove that, we use an approach quite different from the techniques that are used to analyze bubble sort and quicksort. We can tackle the class of all conceivable comparison-based algorithms by modeling every such algorithm as a decision tree. At the root of the tree is an instruction to compare two of the inputs; for instance, it might say "Compare $a[3]$ to $a[7]$". If $a[3]$ is less than $a[7]$, then we proceed to the left branch, but if $a[3]$ is greater than $a[7]$ we take the right branch. Both branches represent sorting problems like the original one, except that some additional information has been gathered about the inputs. We keep on going, making further comparisons until we reach a leaf of the tree; that's when we know the exact ordering of the input numbers. The number of leaves in the sorting problem must be at least $n!$, because that's the number of different orderings the data could be in. Every comparison step divides the number of leaves by at most 2; hence one of the branches of the root must have at least $n!/2$ leaves, and so on. We can conclude that the length of the longest path from the root to a leaf, which is the worst-case number of comparisons needed for sorting, is at least $\lg n!$, which is $n \lg n + O(n)$.

Edgar: When you analyze a specific sorting algorithm, do you also distinguish between different kinds of input? For instance, a list that is almost sorted, or a list that is sorted in the reverse order?

Don: Yes, we need models of the input if we're analyzing the average running time instead of the worst case. Sometimes it's quite difficult to come up with an adequate model. The modeling problem isn't actually much of an issue with respect to sorting, because the difference between the average case and the worst case isn't very dramatic when numbers are being sorted. But when we deal with combinatorial problems, the difference between an average case and a worst case can be twenty orders of magnitude or more. Right now I'm writing about satisfiability, and in most cases I don't know of any good models of the input that match up with practical experience. Some models aren't too bad, but they give insight only into a few small parts of the world of satisfiability problems.

Edgar: Unifying, as opposed to distinguishing between cases, is what theoreticians like to do. Perhaps that's one reason why they call their discipline a "science." Do you then use the word "science" differently than, say, the complexity theorists?

Don: I think science is knowledge that we understand well enough to explain to a computer. That's my definition of science. It has nothing to do with unification.

Edgar: Here are some papers by Hartmanis & Stearns [39], Cobham [10], and Edmonds [24], published in the first half of the 60s.

Don: I don't know this paper of Cobham at all. He evidently wrote it in the days before we had the notions of P and NP. We had questions of solvability versus unsolvability instead. Edmonds pioneered the idea of what we call a good algorithm, an algorithm that always runs in polynomial time. And that idea was extended to nondeterministic polynomial time by Cook about five years later [13]. This 1959 paper by Rabin & Scott [101] was about solvable and unsolvable problems related to finite state machines.

Edgar: What about pointer machines? At the end of Volume 1 of TAOCP [75, p.463], you wrote about defining "a new kind of abstract machine or 'automaton' that deals with linked structures," also called a pointer machine. Then you made a connection between pointer machines and asymptotic methods.

Don: The only data structure in computer programs of the early 1950s was an array. In this section on history from which you're quoting [75, p.457–465], I mention on page 458 that Newell, Shaw, and Simon introduced list processing techniques while studying artificial intelligence: They realized that linked lists, not just arrays, are useful inside the computer too. Some other people picked up on this, and wrote libraries of subroutines that provided operations on linked lists. The concept of a doubly linked list and similar structures arose in the early 1960s. But the general opinion in those days was that, if you're going to write a program that makes use of such data structures, then you ought to use somebody else's library. You wouldn't think of actually writing your own subroutines to insert data into a linked list, nor would you realize that such an operation is simple enough to be done directly within a step of your algorithm. You'd think of list processing as some kind of prepackaged deal. In fact, people usually wouldn't use lists at all unless they wrote their programs in LISP or IPL. These programming languages, as opposed to ALGOL and FORTRAN, specifically dealt with structured data.

I learned about the utility of linked-list structures by reading the compiler for Burroughs Algol (also called BALGOL). Just before I left Case I had run across that amazing compiler, which was much faster than anything else I'd ever seen.[4] Not only was it fast, it also produced well-optimized code. So my first priority, upon arriving in California in the fall of 1960, was to go to Burroughs and meet the people who had built BALGOL. Joel Erdwinn, the principal architect of BALGOL, had left to join Computer Sciences Corporation; but the rest of the team was still there. To my delight, they hired me as a consultant and gave me the job of documenting that compiler, by going through the assembly listing and inserting comments to

---

[4]Note by Knuth after returning home: I just ran across an article by Robert Braden [6] that has a lot of info about BALGOL. It's an excellent example of the kind of history writing I like best, although not done by a professional historian!

explain why it worked so well. My comments were then punched onto the cards and published with the compiler listing. People at Stanford told me later that they had received a listing of BALGOL in 1963 or 1964, with my comments on each line of code, and it had served as their "textbook" about how to write a compiler.

The big breakthrough in BALGOL was due to using linked lists inside the compiler, everywhere. They had 20 kinds of lists, and those lists were all sharing memory. It was easy to insert and delete list items without using a special package. In those days such ideas were sort of out there on the fringes, known to only a few. My chapter on data structures in Volume 1 of TAOCP had the effect of bringing this revolution to the attention of the world. Drafts of my book were circulated for a couple of years before the book came out. People got excited about the data structures because they were learning a new message about how to represent data.

Then, as I was finishing the volume, I started thinking: Hmm, all the literature is counting complexity in terms of array indexing; they're not taking pointers into account.

Edgar: Right, so you came up with the idea of using a pointer machine (or a linking automaton) as a model for analyzing the efficiency of algorithms. Here's the rest of your quote on pointer machines. You started with a critique on Turing machines and then introduced the idea of a linking-automaton model for complexity theory.

> At the time this chapter was first written, several interesting results of this kind had been obtained (notably by J. Hartmanis and R. E. Stearns) but only for special classes of Turing machines having multiple tapes and read/write heads. The Turing machine model is comparatively unrealistic, so these results tended to have little to do with practical problems.

> We must admit that, as the number n of nodes created by a linking automaton approaches *infinity*, we don't know

how to build such a device *physically*, since we want the machine operations to take the same amount of time regardless of the size of n; if linking is represented by using addresses as in a computer memory, it is necessary to put a bound on the number of nodes, since the link fields have a fixed size. A multitape Turing machine is therefore a more realistic model when n approaches *infinity*.

Yet it seems reasonable to believe that a linking automaton [...] leads to a more appropriate theory of the complexity of algorithms than Turing machines do, even when *asymptotic* formulas for large n are considered, because the theory is more likely to be relevant for *practical values* of n.

Furthermore when n gets bigger than 10^{30} or so, not even a one-tape Turing machine is realistic: It could never be built. Relevance is more important than realism. [75, p.464, Daylight's emphasis]

Don: There was also an issue here of what n meant. When I accessed a pointer, the pointer took some space in memory; but that was only $\lg n$ bits of memory because I was only using part of a computer word. If I was counting the memory usage I counted a pointer access as 1 unit of time. It only would become 2 units if $\lg n$ got bigger than 30 (that is, if n got bigger than 2^{30}), which was bigger than anything we would ever see in those days. But there were other parts of an algorithm, like binary search, where you had to do something $\lg n$ times; that was a different $\lg n$. That was a real $\lg n$. The linking automaton provided an appropriate model of complexity because it gave a cost of 1 in the first case and $\lg n$ in the second. But such a machine was actually buildable in the real world only if n didn't get too large. Random-access memories with a capacity of more than 2^{30} words were unknown in the 1960s; when n exceeded a billion, our programs would have to use tape memory instead of linked lists. Tape memory was analogous to a Turing machine, but Turing machines weren't good models of realistic problems. I preferred a model that matched reality on ordinary sized problems and was utopian with respect to asymptotic problems, rather than a model that would match reality only on huge problems. Furthermore, I observed that when n gets larger yet, like 10^{30} or more, we're leaving the real world completely out of the picture in any case. (A petabyte is only 10^{15} bytes.)

Edgar: In the 1960s you were thus advocating another kind of asymptotic methods.

Don: I wanted something so that people who were competing to get a better algorithm would come up with an algorithm that would not only be better academically but better for the world.

Edgar: I'm afraid that many computer scientists today teach asymptotic methods in a far less critical fashion. They follow an abstract approach, based on Turing machines, in which the practical issues you just raised are ignored.

Don: Right. Students aren't being taught properly. In my article 'The Dangers of Computer-Science Theory' [53], I wrote that you have to understand the assumptions that you're making when you apply a mathematical theory to problems. The assumptions and limitations that are needed to prove basic theorems are often swept under the rug and forgotten. People have tended only to advertise their theorems in dramatic language without making the underlying assumptions clear.

Edgar: Yet, in the 1950s and 1960s, several researchers were trying to step away from the Turing machine model. For example, in 1959, Rabin and Scott wrote that

> The restriction of finiteness appears to give a better approximation to the idea of a physical machine.
>
> Of course, such machines cannot do as much as Turing machines, but the advantage of being able to compute

an arbitrary general recursive function is questionable, since very few of those functions come up in practical applications. [101, p.114]

Hartmanis and Stearns, by contrast, did at some point prefer to measure computational complexity in terms of a Turing machine, albeit a multitape Turing machine. In their words:

This particular abstract model of a computing device is chosen because much of the work in this area is stimulated by the rapidly growing importance of computation through the use of digital computers, and all digital computers in a slightly idealized form belong to the class of multitape Turing machines. [39, p.285]

So that seems to have been a general research agenda: trying to find a suitable mathematical model for a computation carried out on a general purpose computer.[5]

Don: Then there is also the random access model which started to become popular with the 1974 book of Aho, Hopcroft, and Ullman [1]. That model was an alternative to my linking-automaton model. But the random access model encouraged people to think more of matrices and arrays than pointers.

We were using pointers all the time, but not in extremely subtle ways, until Bob Tarjan revolutionized the study of data structures by finding magical new ways to point between parts of data. He introduced beautiful algorithms that were unthinkable before. He put extra invariants in his algorithms that often set everything up in such a way that each step accesses a pointer that just happens to tell you exactly what you need in the next step. He came up with new kinds of pointers that weren't in the original statement of the problem. We were amazed at the key

---

[5]As also conveyed by the historian Mahoney [94, p.133].

ways with which he could manipulate certain parts of the structure while making the pointers easy to update, on the fly, as you walk through them. With his work, beginning about 1972 and gradually getting more and more advanced, the pointer model became significantly more relevant than the random access model. Tarjan was the first to prove lower bounds with the pointer model, in his seminal paper on the Union-Find problem [108]. His pointer machines are in fact more powerful than my linking automata, because he wanted to prove a more powerful lower bound.

Edgar: Aho, Hopcroft, and Ullman discussed the Union-Find problem in their 1974 book [1, Sec 4.7] and referred to Tarjan's forthcoming paper [107], but they did not use the pointer model anywhere in their whole book. Could you elaborate on the Union-Find problem?

Don: To explain the Union-Find problem, assume that we're given n initially unrelated elements; we put each element into a set by itself. Then we take two elements and decide to merge their sets together, saying that they are equivalent to each other; and we might repeatedly merge other pairs. The sets are called equivalence classes. When x becomes equivalent to y, we replace the sets S_x and S_y that currently contain x and y by a single set $S_x \cup S_y$, which is the union of S_x and S_y. That's called a Union operation. There's also a Find operation, with which we can find out whether or not x is already equivalent to y, which is the same as saying that S_x is already the same as S_y. The Union-Find problem consists of a mixture of Union operations and Find operations, in which we want to know which elements are equivalent to each other as a result of merges. (If x is equivalent to y and y is equivalent to z, then x is equivalent to z.)

The simplest approaches to this problem will take n^2 steps in the worst case, but two different ways were known by which the running time could be cut down to $O(n \log n)$. Then Tarjan discovered that, if those two ways were both used simultaneously, the running time could be reduced even further: His method needs at most n times $\alpha(n)$ steps, where $\alpha(n)$ is the inverse Ackermann function. That function, which grows much slower than $\log\log n$ and even more slowly than $\log\log\log\log n$ with any number of logs, is basically the slowest growing function that you could ever imagine. If $\log\log n$ never gets bigger than 6, then this function never gets bigger than 2. It does grow, little by little, as n goes to infinity; it's not bounded by a constant in the ideal world of mathematics.

Tarjan also found a matching lower bound, first in a restricted model [107] and later in his general pointer machine models [108]: He proved that, if you work on a pointer machine, there's no way to solve the Union-Find problem in worst-case linear time (that is, n times a constant as n goes to infinity). You always need to do extra work, measured by n times this very very slowly growing function.

Edgar: So the results don't carry over: A mathematical result for the pointer machine does not necessarily hold for the random access model and vice versa? [Don: Yes.] I believe complexity theorists prefer to stick to multitape Turing machines and other popular models of computation, such as the random access model, because then the mathematical results *do* carry over from one model to another modulo a polynomial run-time penalty.

Don: No; random-access automata and pointer machines *are* polynomially equivalent. They differ only when we make finer distinctions, like between linear time and $n\alpha(n)$. The pointer machine model hasn't become more popular than the RAM model, because complexity theorists are happiest with a model that makes it easiest to prove theorems.

Those guys have a right to study polynomially fuzzy models, because those models identify fundamental aspects of computation. But such models aren't especially relevant to my own work as a programmer. I treat them with respect but I don't spend too much time with them when they're not going to help me with practical problems.

Edgar: What about the community that has organized this week's conference in honor of Flajolet?

Don: This community likes to nail all costs of computation down to the last penny. We want to be accurate. They are nitpickers like me, instead of loose livers. I don't say that other people are sloppy, I'm saying that we are anal retentive, i.e., super accurate. That's what my school likes: We want results that are not only ballpark estimates but right smack on. And we need special mathematics in order to develop those methods.

4. Grammars

Edgar: Although the analysis of algorithms is the main topic of this interview, I would like to revisit briefly the subject of programming languages and compilers, which we discussed during our previous conversation [88].

I read Floyd's 1961 'A descriptive language for symbol manipulation' [25] and the review of his article that you wrote in 1962. In your words:

> An algebraic compiler can be described very precisely and compactly in this notation, and one can design such a compiler in Floyd's form in a few hours. [48]

Don: Although I don't think Floyd produced any compilers himself until later, he clearly foresaw how to do it, and he worked at Computer Associates where several other people like Tom Cheatham were writing compilers. By the way, I'm surprised that you found this quotation from my grad student days; I had totally forgotten about writing this review, which came out several months before I ever met Bob in person (at the ACM conference in Syracuse).

At that time in my life, parsing and context free grammars were the bridge for me between mathematics and computer science. Context free parsing was a mathematical topic and I was a mathematics student, but they also related to my separate life as a programmer. For me, the study of context free languages was the one thing that was sort of equally present in mathematics and in computer science.

The hashing problem that I encountered that same summer (cf. [69]) proved to be a second bridge. It was a case of me, a programmer, solving a mathematical problem about the running time of programs. Mathematicians had studied Queuing Theory, and some sorting algorithms had been analyzed by Demuth and others, but those were about the only domains in which mathematicians had studied the running times of processes.

Independently of all this, a community arose around complexity theory, which was called "switching and automata theory" at the time. Many people with an algorithmic bent realized that lots of beautiful abstract ideas were waiting to be discovered about the costs of computation. They took a different attitude than I did, because they weren't writing as many programs as I was.

Edgar: A common research agenda on grammars in the 1960s seems to have been:

> Searching for a class of grammars that was big enough to describe the syntactic constructs that you were interested in, but yet restricted enough so that you could construct efficient parsers from it.

This is a quote from Aho in an interview with Mahoney.[1]

Don: We had lots of possible models for defining languages. If we made them too general, like context sensitive, then we ran into unsolvable problems; and even for context free grammars, it was unsolvable to decide whether or not a context free grammar is ambiguous. Compile time used to be a significant consideration too: In order to use a compiler in those days, I would feed a deck of cards to the computer and the compiler would read them one by one. After each card had been input, I'd see the machine's lights flashing on the console for awhile until it would punch a

---

[1] Available from Mahoney's website:
www.princeton.edu/~hos/mike/transcripts/aho.htm

few cards of output and continue. The waiting time to get my source code translated into machine language would be twenty minutes or so. Therefore we wanted grammars that could be parsed rapidly.

Edgar: In your 1962 critique, entitled 'Backus' Language,' you referred to so-called "Markov-finite state languages mentioned by Chomsky" and commented on something Saul Gorn had written earlier. In your words:

> Now to show why I think Backus notation is stronger than "regular expressions," etc. I claim [...]

> [T]he class of strings *ab, aabb, aaabbb*, etc. *can* be represented in Backus notation, but *cannot* be represented in regular expressions. [89]

Don: Again, I don't remember this at all! [Don looks at the paper, which was actually a letter to the editor of the ACM *Communications*.] I know that I learned about grammars from Chomsky's book [9], but I don't recall that he called them Markov-finite state languages. This letter was written in January, 1962 — my goodness, Jill and I had just moved into our first house. I probably mailed it a week before I was asked to write *The Art of Computer Programming*! I see here also that I didn't give credit to Mel Conway, who I think was responsible for the language extensions that I simply said were "developed at Case Tech last summer" (see [11]).

I thought I had learned my lesson by then. I didn't know the standard conventions about giving credit to other people in scientific publications, I was only interested in telling a story. My experience was mostly confined to reading popular magazines, which would present facts without giving the sources. .... [Don continues reading the paper.] This is a very interesting artifact of naïvety. I don't think it's in the list of publications in my vitae.[2]

---

[2] Actually it's there, as number Q2 under 'other publications' [83, p.247].

Edgar: It shows that you were coming to grips with Chomsky's work.

Don: Yes, I had gotten back from my honeymoon and I was thinking about context free grammars.

Edgar: In 1964, Floyd's 'Bounded Context Syntactic Analysis' came out. At the end, we get to see a discussion between the researchers [26, p.65–67].

Don: Transcripts of the questions and answers following conference presentations were common in those days.

Edgar: Gorn seems to have been rather active in those discussions.

Don: Well, he was a windbag. I never got much out of anything he said, but he kept saying things. I'm obviously not being fair to him, but I have to confess having a distinct impression that he loved to talk but he didn't really have much to say. I might be totally wrong on this. He was a nice man, but I never thought of him as a scientist.

Edgar: The rest of Aho's retrospective comment is about your work on LR(k) grammars:

> [B]ut Knuth [...] said "This is the natural class of grammars that you've all been groping for. And here is its definition. Not only that, but if you restrict yourself to this class of grammars, then here is the procedure for being able to construct a parser from that grammar."

Don: As you say, we'd all been thinking about how to find a good balance between what you could implement efficiently and what you could define. As I was writing Chapter 10 of TAOCP, which was about parsing, I went through all the literature that I could find, including some PhD theses that had been written about grammars that went beyond bounded context, theses by Bill Lynch[3] [93]

---

[3]Incidentally, Bill worked with me on the FORTRAN compiler for Univac SS90 during the summer of 1962; I was living in Madison on the day I "discovered" the analysis of algorithms.

and a couple of others. I finished Chapter 10 and it was Christmas time. Then it occurred to me that there was something more general waiting to be defined, something that encompassed all the ideas I'd read about and more. So I wrote a long letter to Floyd about these preliminary thoughts, and a few months later I formulated the LR(k) concept in a rather complex paper — having no intention to include such far-out extensions in Chapter 10 of TAOCP. ... It was other people who clarified LR(k). I presented the ideas but I hadn't internalized them yet.

Edgar: I would like to contrast Wirth's 2010 remark, which is about Floyd's work, with your own experience. Wirth says:

> [T]here was this new idea, instead of top-down use bottom-up parsing. I continued working on Floyd's idea of precedence grammars, one of the reasons was that then you would not need recursion: Everything was a table of syntax data and they had an interpreter running on that. But after the ALGOL-W compiler, I abandoned all that work and went back to the recursive descent, which is so much more transparent. [19, p.116]

Don: Recursive descent is top-down. Bottom-up says you start at the bottom and look at the precedences and build up that way. It is true that Wirth's `ALGOL-W` compiler was organized around precedence ideas. When you are using recursive descent, you have a reason for everything you're doing that you plan in advance and you sort of know where you are. If you find something wrong, then you can say "Oh, this is an error, because there were no closing brackets to this **for** loop that the user started." But, if you are going bottom-up and there's no closing bracket, then you just keep plunging ahead, waiting until some delimiter triggers an action; maybe a later parenthesis will close the loop by mistake. The structure materializes as you read it, while with top-down you are anticipating what the structure should be. You feel you are in control, while with bottom-up the user is sort of controlling you. That's what Wirth is explaining. LR is sort of bottom-up; yet it has the property that everything you are doing does appear on the stack, and that stack does retain aspects of recursive descent. I'm not really sure about how to compare error recovery in the two models right now, I've forgotten that.

Edgar: Floyd's 1964 paper, 'The syntax of programming languages—A survey' [27], gives me the impression that researchers were reflecting rather strongly in the early 1960s. I mean, this bird's eye view on compiler technology did not come decades later as Wirth's 2010 comment alone might suggest.

Don: I'm not sure what you mean by "decades later," because Wirth finished ALGOL-W in the 60s and began on Pascal in 1970 or so. At any rate, compiler research was certainly intensive, representing roughly one third of all computer science in the 60s, and Floyd himself wrote many of the most fundamental papers about it. From my point of view, his survey [27] introduced me to the important idea of what I call recursive coroutines. His article described a model that we would now call object oriented. Floyd described parsing in terms of employees of a parsing company. One employee is a specialist in expressions, and he can hire someone whose job is to get terms. The term guy can then hire somebody whose job is to get primary expressions. When they succeed, they report to their supervisor, who then requests something else, and so on. The supervisor might also ask them to try again, if their initial successes don't lead to a consistent parse. These workers communicate by means of a global pointer to the current position in the input string, which is visible to all of them on the wall of the building. This idea of individual actors, working together and hiring each other and reporting back and continuing on, is extremely useful. It goes beyond ordinary subroutines because each agent has its own personal state; the supervisor can say "Okay, you found this term but I don't want that one, find another one." It was a beautiful new paradigm for programming, a new way to organize processes. Bob presented the concept in an anthropomorphic way that made it also understandable.

Edgar: Did people pick up on that idea?

Don: I'm not sure how many other people realized the genius of it, because it was so well done. From that moment on, I started thinking about more ways to apply recursive coroutines. I wrote a paper about the subject as my farewell to Ole-Johan Dahl [92]. Floyd had described the concept beautifully, yet the idea of recursive coroutines did not spread as it should have until SIMULA came along.

5. Algorithmic Barriers Falling

Edgar: Coming now to automata theory, in your joint 1967 paper with Bigelow entitled 'Programming Languages for Automata,' you showed that "techniques of automatic programming are useful for constructive proofs in automata theory" [87].

Don: The people who were writing papers on automata theory were describing everything in terms of tuples of numbers, something like a computer's machine language, without any intuitive structure. They would devise intricate numerical formulas by which one could compute what state follows another. Bigelow was an undergraduate student who worked with me. We asked, "Why don't you have symbolic names for these states, so that you can see what you're doing?" But unfortunately, this paper did not generate any sequels.

Edgar: In the paper you wrote:

> [T]o show that "software" techniques, which have been developed for programming "real" computers, are equally useful for constructing automata programs. [87, p.615]

> [Our] programming language can be applied to automata theory, as we prove there are non-context-sensitive languages accepted by a stack automaton. [87, p.615]

Don: Not only did I want to present a syntax that would yield a higher level language to describe automata, I also wanted to demonstrate its utility by proving a theorem that hadn't been proved before. If I had proved that theorem with the old methods, it would have been very hard to understand my proof. My goal in this paper was to say: "Look, we can come up with better proofs if we describe our automata with a high level language instead of in set theory." Traditional methods of formal mathematics, which gave numbers to the states, led to error-prone constructions, like the mistakes Turing had made in his classic paper that introduced Turing machines. If he had had a high-level language to describe his machines, he wouldn't have made those mistakes. Of course high-level languages were unknown in Turing's time, but by the 1960s such languages were commonplace.

Edgar: Concerning this connection between programming practice and automata theory, you have referred to J. H. Morris and his 1969 text editor for the CDC6400 computer in [90]. In early 1970 you learned about Cook's "surprising theorem about two-way deterministic pushdown automata" [12, 14]; you thoroughly studied it, and subsequently came up with a novel practical algorithm that Morris had invented independently in a different way.

Don: People were discussing a curious, purely academic problem about recognizing sequences of palindromes. A palindrome is a word equal to its reverse; and a "palstar" is a *sequence* of palindromes: one palindrome followed by another palindrome, followed by yet another, and so on. The question was, if you have a sequence of a's and b's, can you recognize whether or not it is a palstar? I could imagine various algorithms for that problem, but they weren't guaranteed to be fast.

Steve Cook had solved that problem in a novel way by introducing what he called a two-way deterministic pushdown automaton, about which he proved a remarkable theorem: If you could program his automaton to recognize any set of strings, then your program could be transformed into another program for a random access machine, which would recognize the same set of strings in linear time — no matter how fast or slow your original program was when it ran on the automaton itself! Furthermore, Cook's automaton could recognize palstars. Therefore an algorithm exists, which recognizes palstars in linear time on a conventional computer.

I was dumbfounded, because I thought of myself as a good programmer; yet an automata theorist had established the existence of a program that was faster than any that I could think of. I looked at his theorem and said to myself, "How am I going to do such a thing? This is a miracle." It was the first time that automata theory had taught me something that I wasn't able to discover with ordinary common sense.

Okay, Steve showed that a linear-time algorithm exists; but what was it? On a big blackboard, I wrote out a palstar-recognition program for Cook's machine, using states and tuples. Then I hand-simulated his method of converting such a program into a sequence of steps for a random access computer. All of a sudden I had an "aha" moment, realizing what I had missed when I had tried to devise my own algorithm.

Do you see what I mean? By carrying out Cook's mechanical transformation process, as applied to the state transitions for a palstar recognizer, I was finally able to psych out what its actions meant in the context of palindromes.

After that "aha" moment, I took up another problem that Cook's machine was able to do: pattern matching. Again I set up the tuples and applied his transformation and simplified the results. And this led to a very efficient algorithm for pattern matching that was, in retrospect, quite obvious in some ways although it hadn't been apparent to me.

Jim Morris had already discovered this method on his own, without relying on automata theory. Vaughan Pratt had also come up with an equivalent algorithm, I don't know how. Dick Karp mentioned their work in the spring of 1972 (see [98, p.176]). Morris, Pratt, and I eventually learned of each other's independent routes to discovery, and I wrote the results up for publication as a joint paper [90]. The original title of my paper was going to be 'Automata Theory Can Be Useful,' because it illustrated how our minds could be blocked by a certain way of thinking but then the automata theory reveals a fruitful new direction.

This process of deciphering low-level machine code and converting it into understandable high-level concepts has only been successful for me a few times, and I wonder if tools could be implemented that would facilitate such activities. My PhD thesis [50] was another example of a fortunate "aha." The computer had output two sets of binary patterns, each of which contained 32 entries. I understood one of the entries in one of the tables, because it represented a finite field. What I needed to do, in order to make sense of all the output, was to find a mapping between the 32 patterns on the left and the 32 patterns on the right. If I could find a nice correspondence between these two sets, then I would probably be able to apply that correspondence to other problems: I'd know the structure of the guys on the right, because I knew the structure of the guys on the left. I was lucky again, because I did find such a correspondence, and it became the basis of my PhD thesis because it resolved a longstanding conjecture in combinatorics.

Edgar: Were you at first very skeptical about the usefulness of automata theory and of Cook's theoretical result in particular?

Don: Before I saw Cook's paper I was skeptical that automata theory would prove anything that I wouldn't have already known somehow. I regarded it as a nice abstraction for thinking about algorithms and for proving things impossible, but not for proving things possible.

Edgar: When did you meet Cook for the first time? Did you get to know him well?

Don: Shortly after beginning to work at Stanford, I crossed the bay to visit Dick Karp in Berkeley, and I ran into Steve Cook at the same time. We talked briefly but I didn't get to know him especially well.

Incidentally, I met Vaughan Pratt on that same day. Karp and I were having lunch, talking about everything under the sun, and Pratt came to sit at the same table. I was working on various problems related to Volume 3 of TAOCP. Vaughan sat in on the conversation; he was a new graduate student at UC Berkeley. I told Karp about four research problems that I was currently stumped on; Vaughan heard what I said, and a week later he mailed me solutions to two of them! That was when I said: "Vaughan, you've got to come to Stanford. How did you do these things? I'd love to work with you some more." He came to Stanford soon after, and finished his PhD thesis in a two-year record time.

Edgar: I would characterize Stephen Cook as a logician. Is that fair?

Don: He has a profound understanding of logic and its connections with algorithms. The Berkeley faculty made a terrible mistake when they decided not to give him tenure. He moved to Toronto in 1972 and gave lectures there, which demonstrate his deep intuition about algorithms. I know this because I'm writing now about satisfiability, which is the topic of Section 7.2.2.2 of TAOCP, and I've seen notes from some of the lectures he gave at that time. My book will publish for the first time an algorithm that he presented to his class in January of 1972, an elegant way to solve satisfiability problems that has turned out to be prophetic. He called it 'Method I,' and I'm calling it 'Algorithm I'; modern SAT-solvers are based on ideas that are very close to what Cook told his class. He didn't foresee the entire picture, but he introduced a brand new way of attacking the satisfiability problem that was forgotten for 50 years.

Edgar: What about Karp? Was he a logician?

Don: No, I think of Dick as a dyed-in-the-wool computer scientist, essentially in the center of our subject — even though I guess he doesn't actually write many programs. He has always been among the leaders of computer science, especially with respect to theoretical issues. His background was in operations research rather than in logic, and he worked for IBM during most of the 1960s. He is a master of many different things, but especially combinatorial algorithms. In the early days he published seminal papers on information theory and codes, on the Traveling Salesrep Problem, and on new models of automata that were actually practical. Now that you've asked the question, I can't help wondering how Karp would have defined *himself*, if he had been asked in 1967 (as I was) about what he considered to be his life's work.

Immediately after Cook came up with the idea of what we now call NP-completeness [13], Karp showed that it was really a big thing [43], because he demonstrated that more than a dozen really important problems were all closely related to what Cook had just done. Thousands of our colleagues had been working on such problems for years.

Edgar: Did you read these papers by Cook and Karp at the time?

Don: I was running a weekly seminar at my house, a kind of salon. It was called a combinatorial seminar. In retrospect, that seminar has turned out to be very nice from an historical standpoint, because every week I asked every attendee to sign the guest book. Now we can look at that book and see exactly who was attending each seminar. Karp spoke at one of the first meetings, on 6 December

1971, and forty others were in attendance. This was several months before he gave his conference presentation [43], so we heard about it right when it was new.

Edgar: What about Cook's 1971 paper, 'The Complexity of Theorem-Proving Procedures' [13]?

Don: Cook's work had come out in a conference that I didn't attend, but Dick did. He came back with the news. The full story is, of course, told here in another book that you've brought with you and put on the table. [Don points to Garey & Johnson's book [30].] Meanwhile Jill and I were getting ready to spend a sabbatical year in Norway, where we arrived in June of 1972.

Edgar: Karp showed that many famous problems — from graph theory, computational logic, switching theory, combinatorics, and mathematical programming — were essentially equivalent to the satisfiability problem, which in turn had been cast as a language recognition problem.

Don: Yes, and problems from other fields including biology came later on.

Edgar: Did Karp present a fundamentally new insight? Were Cook and Karp's results interpreted optimistically? Or were these problems perceived as too difficult to tackle efficiently on a real computer?

Don: Right after Cook's paper there was a brief period of optimism, because we all had thought that satisfiability was a trivial problem. People had said "Hey, to decide whether or not a given Boolean function is always equal to zero, what could be simpler than that?" Now Cook and Karp demonstrated that that problem packed a wallop: If you could solve it efficiently, you could also knock down all these other problems as well. Cook himself had come up with a new algorithm, Method I, whose efficiency was unknown. We thought that the task of taming the satisfiability problem wouldn't be especially hard; and our optimism lasted about three or four months. But then we realized that our methods were actually pretty slow when we tried to scale them up.

Another kind of optimism was actually in the air, and that optimism was really the reason for the conference in March of 1972 where Karp gave his presentation. Almost all of the classical algorithms we knew were being replaced by faster methods. Matrix multiplication, for example, went down from n^3 steps of computation to only $n^{2.8}$. Algorithms that had taken n^2 time were now reduced to $n \log n$. New algorithms were being invented all the time. People discovered a linear-time way to calculate the median of n elements. Every week we'd hear of another problem that had been given a better algorithm.

Edgar: A better algorithm in terms of worst-case complexity?

Don: Yes, worst-case complexity. But still the algorithms were good enough that we would actually program them. They weren't only asymptotically good, they were faster in practice. The fact that nonobvious algorithms exist, coupled with Tarjan's brand new ways of using pointers to obtain linear-time algorithms, was just mind boggling. We didn't yet think much about Moore's law regarding advances in hardware; we were dramatically improving the speed of software. That was the flavor of the early 1970s. Algorithmic barriers kept falling and falling and falling.

I soon began to realize that I wouldn't be able to finish Volume 4 nearly as soon as I had thought, because all these algorithms were being invented just about the time I finished Volume 3. That was in the spring of 1972: I wrote the hashing section in February and the final section on multidimensional searching in March. In April I wrote Forsythe's obituary [55] as well as my paper about Babylonian algorithms [54]. In May we packed up and got ready to go to Norway. Eventually this book by Garey and Johnson [30] came out, greatly extending the lists of problems that are NP-hard and NP-complete.

In our previous meeting I told you that 1967 was the busiest year of my life [88]. Well, the first part of 1972 was the second most hectic time in my life, because I was finishing Volume 3 and hosting seminars every week, and learning about all kinds of new algorithms coming out. Then I went to Norway and spent the year reading papers by Edmonds [24] and others on efficient algorithms, learning that the task of writing about combinatorial algorithms was going to be quite a challenge. If you look at the computer science journals in 1974, 1975, 1976, 1977, you'll see that more than half of the articles were about combinatorial algorithms — things that had to be discussed in Volume 4. So the whole subject was fermenting and boiling over. I kept making draft notes all during this period, and started to make what I thought would be the final push towards completing that volume in the spring of 1977. By April of that year I'd typed about 100 pages, covering Boolean functions and bitwise operations. But then my work on combinatorial algorithms was interrupted, because I suddenly realized that I must take a detour into typography.

Edgar shows Don two of his articles:

- 'A Terminological Proposal,' SIGACT News, January 1974 [57].

- 'Postscript About NP-Hard Problems,' SIGACT News, April 1974 [60].

Don: Yes; I can explain more of the background. Jill and I were in Norway for 15 months, having left California at the beginning of summer 1972 and returning at the end of summer 1973. During the first summer we took language lessons, and in the second summer we were vacationing on an island called Tjome. During the academic year I gave lectures at Oslo University, every Wednesday for two hours, and the notes were written up each time by a different person in the class. Those notes are available through the University of Oslo archives.

The Oslo notes cover the algorithms that I was learning at the time; and five of the lectures, from 31 January through 28 February of 1973, were devoted to what I called 'Hard Problems.' Theorists were calling them "polynomially complete" problems, but I didn't like that name; it just didn't sound right for such an important concept. It was going to confuse people. On the other hand I didn't like the term "hard problems" either. My lectures, incidentally, proved Cook's theorem in a different way, by using Markov algorithms instead of Turing machines as the underlying computational model.

Soon after returning to California I sent letters to several dozen people, asking for advice about how to change the name of Cook's hard problems from "polynomial complete" to something more suitable. I was toying with alternatives, like "Herculean" or "formidable" or "arduous"; see [57, p.12]. None of them actually worked well. But many of my correspondents independently came up with the excellent terms NP-hard and NP-complete, which I'm glad to say were quickly adopted. This was one of the most fruitful episodes I've ever come across, in the history of naming new concepts, and I was pleased to serve as a catalyst for that change.

Incidentally, I received galley proofs from Aho, Hopcroft, and Ullman of their great new book [1], in November of 1973. Their Chapter 10 was devoted to "polynomially complete problems"; so I wrote to them early in December, begging for a change in that terminology before it was too late. They replied that, alas, it *was* too late to make a change, because a great many pages were affected. Fortunately, however, one of them (I think it was Jeff Ullman) decided to bite the bullet and change every related page, because I was able to show him the results of my poll (cf. [57, p.13]) just in time.

Edgar: So you wanted to introduce new terminology that would be appropriate for the general public. One of your proposals was to use the word "formidable." And here's an excerpt from 'A Terminological Proposal' in which we see Pratt respond to your suggestions.

> One final criticism [...] was stated nicely by Vaughan Pratt:
>
>> "If the Martians know that $P = NP$ for Turing Machines and they kidnap me, I would lose face calling these problems 'formidable'."
>
> Yes; if $P = NP$, there's no need for any term at all. But I'm willing to risk such an embarrassment, and in fact I'm willing to give a prize of one live turkey to the first person who proves that $P = NP$. [57, p.18]

Don: You have to know the background of that in-joke. In Poland, Stanisław Mazur had proposed a problem in 1936 about the representation of continuous functions, promising to reward the solver with a live goose. The problem became famous, and it remained open until 1972, when a young Swedish mathematician named Per Enflo finally resolved it. Mazur duly presented him with a live goose, in a well-publicized ceremony, and all theoreticians in 1974 were aware of this story.

Incidentally, Per Enflo was also a world-class pianist. He was visiting Stanford when I returned from Norway, and I had a great time sightreading piano duets with him on 14 December 1973.

Edgar: At this stage in your career you thought that P is not equal to NP and I suppose most people thought so. [Don: Yes, after the summer of 1972.] We discussed a couple of minutes ago that, for the first three to four months, everybody thought P and NP denote the same class of algorithms.

Don: Well, not everybody, but optimism was certainly rampant. All we had to do was solve a simple problem known as satisfiability.

Edgar: What's the relation of all this with the 'Concrete Mathematics' course that you introduced at Stanford in 1974?

Don: Concrete mathematics is an independent topic, related to the analysis of particular algorithms rather than to computational complexity. While analyzing algorithms and writing my book drafts, I found myself using various kinds of mathematics that I had never been taught. I had pretty much seen them only in the problem pages of mathematical journals. Somebody would ask for the explanation of a mathematical pattern, and the solvers would reply by using techniques that turned out to be really fundamental for the analysis of algorithms. In TAOCP I tried to organize these techniques into a logical sequence that students could learn. Meanwhile, I was in a mathematics department, and this style of mathematics was actually not at all in fashion. Bourbaki was king: The more abstract you could be, expressing everything in terms of morphisms and categories, the better. Highly abstract methods were in favor in all the best mathematical schools. In more and more of the lectures that I was hearing at Caltech, I would find myself sitting in the audience saying to myself, "So what? So what?" Eventually I switched fields and became a professor of computer science, where I could deal every day with concepts that truly excited me. As a math professor I'd been getting less and less enamored of the trends in pure mathematics.

Of course I wasn't the only mathematician who was unhappy about those trends. People viewed the situation as a tension between concrete and abstract. There was tension between solving problems that involve actual numbers, versus solving problems that use only symbols and words and are generalizations of generalizations. I'm exaggerating, but there was an editorial in the *Notices of the American Mathematical Society* [105] entitled 'Can Mathematics Be Saved?' The author said that this overly abstract trend was taking us away from our roots so that we no longer could solve real problems, only academic ones. A particularly thought-provoking paper was written by John Hammersley in 1968, entitled 'On the enfeeblement of mathematical skills by "Modern Mathematics" and by similar soft intellectual trash in schools and universities' [38]. (I recently quoted from this wonderful paper in response to Bob Sedgewick's question about whether everybody should be a programmer [110].)

So I was reacting against all that; I wanted to teach computer science students at Stanford how to analyze algorithms. They were going to need to know more than Bourbaki had to offer. I introduced this course and called it 'Concrete Mathematics,' as an implied antithesis of Abstract Mathematics. I taught it in the computer science department without permission of the mathematics department. But I had students from the mathematics department, including some like Peter Sarnak who have turned out to become leading mathematicians of today. Not only that, I had great computer science students from France, who turned out to be the best students in 'Concrete Mathematics' in spite of having been trained Bourbaki-wise at the *grandes écoles* of Paris! I taught that class every year or every two years — a dozen times between 1970 and 1986, and I completed the book at the end of 1988.

Each year we had class notes; the students and I would discuss topics during class, and the teaching assistants would write down what was said. (Vaughan Pratt was the teaching assistant in 1970; Oren Patashnik in 1981 and 1984.) So I could use all of those discussions and put them together into a book. And the book actually serves nicely as a supplement to Volume 1 of TAOCP, because it teaches basic techniques that are useful when you're doing analysis of algorithms, among other things. The book is also a sort of manifesto, because it champions the way that I like to do mathematics.

Edgar: In the beginning of this discussion, we distinguished between your take on asymptotic methods and the large literature in which people play asymptotic games.

Now you're contrasting your analysis of algorithms with Bourbaki's work. That's a different kind of contrast. What was Bourbaki abstracting away from?

Don: Bourbaki was saying: We've got rings and fields, but these are special cases of modules, which are generalizations of vector spaces, or something like that, which really is a finite dimensional case of something else, and so on. The really important problems for Bourbaki dealt with infinite systems. He presented a grand vision of large parts of mathematics. It has great beauty, but it neglects specific examples.

At least, that was my impression of the Bourbaki volumes at the time. I must confess that I hadn't actually *read* Bourbaki when I introduced Concrete Mathematics at Stanford, except for its historical sections. I mentioned a minute ago being surprised to find that the mathematical skills of students from France had *not* actually been "enfeebled" by their abstract upbringing.

Edgar: These mathematicians were emphasizing the infinite, while you wanted to focus on the finite?

Don: Right. They have a wonderful point of view, but you've got to build up to it instead of come down from it. So it didn't help me analyze any algorithms. To analyze algorithms I had to know how to compute sums of series. Bourbaki's volumes didn't deal with that, they would just say that in principle such things can be done. I needed a good way to say that one particular series is summable in terms of familiar functions and another series is not. And I wanted to stress the importance of special sequences of integers, such as the two kinds of Stirling numbers.

I retrofitted the word "concrete" in 'Concrete Mathematics,' by saying that it's a blending of continuous and discrete mathematics. But that idea wasn't really in my mind when I started the course. The essence of my book can probably be described more accurately by calling it "manipulative mathematics," because the main ideas tell you how to transform one formula into another, and how to think about formulas as the reality instead of numbers. Students of my course became very good at saying "This formula equals this formula, and I can simplify that sum by changing the formula into another one where the variable of summation only occurs once instead of three times." Furthermore, when the variable of summation occurs only once, they could often sum it and obtain a simple answer. Learning how to manipulate formulas fluently, and learning how to see patterns in formulas instead of patterns in numbers — that's what my book *Concrete Mathematics* [37] is essentially about.

Edgar: Which was also the topic of Manuel Kauers this morning?

Don: Right. In fact, he and Peter Paule in Austria recently published a beautiful book called *The Concrete Tetrahedron* [44], which is sort of the sequel to *Concrete Mathematics* [37].

Maybe we should go and have something to eat.

6. Intermezzo

Don and Edgar move to the apartment's kitchen for lunch.

Don: Another thing that I forgot to mention in my talk on historiography [86] was an example or two of history that I think is being done right. I asserted that recent issues of the journal *Isis*[1] seem to be straying too far away from internal history; but I forgot to point out that I think the journal *Notes and Records of the Royal Society*[2] has a really nice balance right now. I happened to learn about it because I was elected to the Royal Society as a foreign member. I began to like the journal so much that I decided to try to meet its editor, Robert Fox, on my next visit to Oxford. He's been retired for some years, but he was the head of the History of Science group there. He's still active as editor, although I think he's looking for a successor. Therefore, last week I invited him to lunch at my college in Oxford and we had a nice time. We found that we have many interests in common. I thanked him for the nice blend of internal and external history in his journal, and he said, "Well, you know, recent authors aren't afraid to teach some science as well as merely talking about the circumstances of a person's life or some of the people they met and so on; they actually say something about biology and what was actually seen in the microscope, instead of how much the microscope cost. And that's because they know they have a captive audience of scientists." Because I am a scientist myself I can't speak for how such articles

---

[1]See *www.press.uchicago.edu/ucp/journals/journal/isis.html*
[2]See *rsnr.royalsocietypublishing.org*

would appeal to the public at large. But I do think that it must be possible to reach a general educated audience the way *Scientific American* does, so I really hope to convince computer science historians to revert to their former ways.

In the 1960s I started to learn a little bit more of the world, and at Caltech I regularly saw *The Illustrated London Times*, a newspaper that would often feature items such as two-page spreads devoted to detailed stories about archaeologists, explaining exactly what they were digging up, as well as other articles devoted to scientific matters. *The New York Times* occasionally included articles like that, but such practices were rare in America. Thus I learned that some British newspapers were written for the upper classes, and that they were markedly different from the British tabloids. In America everything was homogenized, oriented towards the common denominator, so you would almost never find a subject that was treated in any depth whatsoever. I was drawn to these papers from Britain, because they dared to try to teach me something. In a way, however, I felt guilty about reading them because, hmm, the British are elitists. I mention this today because it seems similar to the idea of dumbing down science, as in so much of today's literature on the history of computing. I don't want a steady diet of material intended only for a mass audience, material that actually underestimates the intelligence of mass audiences.

Edgar: Perhaps many historians, themselves, don't really have the background to give a technical account?

Don: Yes; it's absolutely true that historians as well as other scientists aren't going to be the best persons to describe the developments of a field. Martin Gardner was a great expositor of mathematics because he was careful to say that he wasn't a mathematician. He had a large network of correspondents, who would advise him and tell him how to get rid of false chords in the drafts he sent to them. His work was cooperative, it wasn't produced by himself working alone. Ideally we have a team, in which one person understands the readers very well and another person understands the science very well. Cooperation between two people with complementary talents is I think a really powerful way to proceed. It's certainly true that a person can be so inbred that he or she doesn't realize what other people don't know. One of Gardner's secrets was that he, as more or less an amateur, could appreciate what was difficult to learn but also what was beautiful and surprising about the subjects he wrote about.

Edgar: Cooperation between a computer scientist and a trained historian seems the right thing to do.

There are many historians who have interviewed computer scientists on an informal basis. Unfortunately, a lot of these unauthorized interviews will get lost, even though a lot of effort has been made. A similar remark holds for archival material: The writings of a computer scientist get lost once the actor has died, if not earlier. The Dutch professor Jan Willem Klop, by contrast, has put much of his archival material online.[3]

Don: Feigenbaum is doing this at Stanford and has gotten some computer people to make it well-indexed, searchable, and things like that.[4] This guy Chuck Weiner I was telling about, he told stories of visiting the widows of physicists in Florida, finding that file drawers full of papers were being destroyed by insects at the moment that he entered the house. He had to detoxify the scraps that could be salvaged.

Edgar: You referred to Ole-Johan Dahl earlier today. Did you meet him in Norway?

Don: The first time I went to Norway was in 1967, for the IFIP conference in Oslo on simulation languages. It was a major conference that brought many people together for the first time. I had previously served as editor of a paper

---

[3]See *janwillemklop.nl/Jan_Willem_Klop/*
[4]See *library.stanford.edu/collections/edward-feigenbaum-papers*

about SIMULA that Dahl and Nygaard had submitted to the *Communications of the ACM* [17]. I was blown away by that paper, it was such a great advance. So I decided to go to Norway for this conference. I was really impressed by Dahl's deep understanding of issues of programming. He was a very shy and modest person. He had written superb class notes for his students in Norway, unknown in other countries. (I learned to read Norwegian afterward. It's a little bit like Dutch.) At this conference he brought with him some music to play piano duets. I loved to sight-read music, and he was a fantastic musician. He introduced me to a whole tradition of music that was a revelation to me. That became quite a bond between us. Every time we would see each other, years later, we would spend some hours playing four-hands piano music.

Dahl was the main genius behind SIMULA and object-oriented languages. Nygaard was strong too, but he was oriented more towards salesmanship and to getting a job finished. Dahl was an introvert and Nygaard was an extrovert, a politician active in parliament and in labor movements, promoting good causes, and so on. The deep thoughts were Dahl's. Earlier today I spoke about Floyd's recursive coroutines; Dahl had independently understood all of this too. He incorporated those concepts into SIMULA.

Edgar: I read your letter of recommendation for Floyd in [79]. What I find interesting is that to express your appreciation for Floyd, you put him alongside Dijkstra and described the latter as being "responsible for more landmark contributions in nontheoretical aspects of computer science than any other man" [79, p.8].

Don: Dijkstra saw things through a programmer's lens, with an emphasis of quality over expediency. His famous early algorithms to find shortest paths and spanning trees [21] were published before his work on ALGOL. And he also had very broad interests, as I said earlier. For instance, his

EWD notes[5] refer to a lot of geometry problems and other aspects of pure mathematics that aren't necessarily related to programming.

The paradox for me is that Dijkstra stopped writing computer programs! Instead, he tried to find sort of the essence of ways to present mathematics.

Edgar: Working with real machines is what you, Dahl, Dijkstra, Hoare, Naur, and Wirth had in common in the 1950s and 1960s. Perhaps the logicians, the pure theorists, needed an academic home.

Don: One of the questions that came up during my talk last night was: "Is anybody in the 21st century going to come up with something that is as important as Gödel's Incompleteness Theorem?" It was a weird question, because I love Gödel's Incompleteness Theorem, but I wouldn't consider it the number-one result of the 20th century. To some people it is a really glorious thing. I don't like to trade one person's aesthetics against another and say that one is better than the other, but I know what I like.

One of the people in the audience last night, a young man, said that he had been a student learning about computing for two years but not really being much interested in it. Then one day he learned the Knuth-Morris-Pratt algorithm for pattern matching [90]; he was entranced by it and he decided that he loved computer science. For him, there was something magical about that idea. Isn't it remarkable? A beautiful algorithm can change somebody's career.

Don and Edgar move back to the living room.

---

[5]See *www.cs.utexas.edu/users/EWD/*

7. More about the 1970s

Right after lunch, Don and Edgar discuss the 1970s again, chronologically, based on some of Don's writings.

Edgar: Going back to FORTRAN, in 1970 you wrote a report entitled 'An empirical study of FORTRAN programs' [51]. One of the things you said is:

> There has been a long history of optimizing the wrong things, [. . .] [51][80, p.508]

> Frequency counts were commonly studied in the early days of computers [. . .], and they are now experiencing a long-overdue revival. [51][80, p.514]

I also find this kind of comment in your 1974 Turing Award lecture:

> [M]any people nowadays are condemning program efficiency, telling us that it is in bad taste. [58, p.671]

It seems like you were countering a general movement.

Don: I was trying on the one hand to counter a trend against caring about program efficiency and, on the other hand, to advertise a healthy new trend towards learning more about the actual costs of program execution. The FORTRAN paper promoted recently developed tools to find the "profiles" of programs, the number of times each part of a program was actually executed. Profiling has actually gone up and down in popularity over the years; in 1970 it was going up. People have revived it periodically, and it went into the design of the Alpha chip processor. That computer (see [4]) included advanced hardware support for gathering runtime statistics, almost for free, to the people who were studying bottlenecks in their programs. Special internal registers could be dedicated to the task of counting how many instructions you had performed. Users could specify the kind of instructions that they wanted to count, making different tests on different days with different subsets of their instructions. Those counts were made directly by the hardware. Such statistics were obtainable in the 1970s only via software; even so, we found them indispensable for tuning up programs that were being used heavily. In Volume 1 of TAOCP I had been careful to annotate each line of each program with an indication of how often that line of code was executed. This practice may have influenced a general interest in profiling during the 1970s.

Edgar: In your 1971 paper 'The Dangers of Computer-Science Theory,' you wrote:

> For many years the theory of automata was developing rapidly and solving problems which were ostensibly related to computers; but real computers could not care less about the automaton theorems because Turing machines were so different from real machines.

> However, one result was highly touted as the first contribution of automata theory to real programming, an efficient algorithm that was discovered first by the theoreticians, namely, the Hennie-Stearns (1966) construction [...] Well, once again the theory did not work in practice; [...] a single tape is only 2400' long; so the asymptotic formulas do not tell the story. [53, p.192]

> When the Hennie-Stearns method is actually applied to a tape full of data, almost 40 hours are required, compared

with only about 8 hours for the asymptotically slow method. [53, p.193]

Don: That method cleverly packed things into two tapes while respecting the strong limitations of sequential tape motion. I tried to program it for a real computer, possibly for the first time.

My paper tried to make two points: On the one hand, theory brings us all kinds of beautiful things; on the other hand, misunderstanding of theory causes harm. So we should learn how to live properly in the presence of theoretical results. This has remained a common problem: People repeatedly apply known theorems without being aware of the limitations of those theorems.

All through science, experience shows that people start to believe certain hypotheses about science that are consistent with the results of many experiments, but they fail to realize that other hypotheses might have been equally consistent. For instance, in biology, if you don't remember which genes you were studying to get a certain behavior, then you might come up with a different kind of treatment for cancer. The people who use scientific results have to be aware of the history of those results, as to who did it and when and with what equipment, and so on. That's where history comes into critical importance, not just for the uses that I mentioned earlier for the advancement of knowledge but actually to prevent bad errors. I guess that's a recurring theme: I'm searching for the best ways to combine mathematical theories with the real world.

Edgar: Were you alone in voicing such critical remarks?

Don: It could have been. On the other hand, there wasn't anybody saying I was wrong. You might be interested in something that happened when I presented this paper in Bucharest. The occasion was an international "congress for logic, methodology, and philosophy of science," and it was one of few conferences to be held behind the Iron Curtain.

Nicolae Ceauşescu was a principal sponsor; he gave the opening speech and came to the conference dinner. I shook his hand and people were complimenting him on the ways in which he was promoting science in Romania. Romania was indeed going through a remarkable transformation; but it was very peculiar, because people didn't have access to libraries, for example, so they couldn't really do science. Every city block in Bucharest seemed to have a bookstore where you could find popularizations of scientific results, but to do new science their scientists were basically shackled. (I knew this because I visited several of them in their offices.) So it was a very peculiar situation. I gave this tongue-in-cheek talk about why we should *not* have theory because it is misused. Marcel Schützenberger was sitting with me — he's the French genius behind combinatorics — and he said "Well, Don, that was a brilliant talk, but it is maybe given in the wrong place. The Party is now going to decide not to fund computer science any more."

Edgar: In Frankfurt you mentioned your dislike for philosophy [88].

Don: "Dislike" isn't the right word, but I certainly don't worship philosophers. You can be a philosopher without knowing anything; you just have to be witty. Anybody can be a philosopher by choosing n philosophers with which you agree, and you also choose n philosophers with whom you disagree; then you can write a paper.

Edgar: Nevertheless, you did refer to the philosophy of William James in 'The Dangers of Computer-Science Theory' in the following manner:

> At this point I would like to quote from some lectures on *Pragmatism* (Chapter 2) given by the philosopher William James at the beginning of this century:

> > When the first mathematical, logical, and natural uniformities, the first *laws* were

> discovered, men were so carried away by
> the clearness, beauty and simplification
> that resulted, that they believed them-
> selves to have deciphered authentically
> the eternal thoughts of the Almighty.
>
> You see that computer science has been subject to a
> recurring problem: Some theory is developed which
> is very *beautiful*, and too often it is therefore thought to
> be *relevant*. [53, p.194, Knuth's emphasis]

Perhaps this connection with William James came via Peter Naur (cf. [18]).

Don: I certainly share with Peter Naur that I am inspired by trying to correct other people's mistakes.

It is also true that many people don't realize how much mathematical rigor *is* needed in practice. The question keeps coming up in our discussion: What kind of theory should programmers know in order to write reliable programs? New applications always ask for another and new kind of rigorous theory. Some people look at programming and say "Well, there's no need for mathematics any more, all we're doing is writing apps." They don't realize how much mathematical rigor is needed in order to avoid race conditions or when you're doing distributed programming that you don't lose control, or whether the kernel drivers aren't missing cycles. So all the things that go on in Google searches, most of them require some pretty good mathematics behind them. Theory is continually relevant, even though a lot of programmers also can do their work competently without much theory. Today there are even more programmers who do need the theory.

Edgar: In 1971 you wrote a paper on optimum binary search trees in which you took the access frequencies into account. In your words:

> [W]hen some names are known to be much more likely to occur than others, the best possible tree will not necessarily be balanced. [52, p.14]

So you distinguished between different frequencies; that is, different ways in which a binary tree is used. How far did you go in making these case distinctions? I mean, at some point you have to generalize as well, right?

Don: If you have a fixed dictionary, then you want to tune a search method to that dictionary. And if you also have information about how often a user is going to look for each word, you can tune your algorithm even better. I've got an obsessive/compulsive nature, so when I look at a problem I think it's the only problem in the world and I want to learn as much as I can about it. If I'm designing search trees, I want to know what is the best I can ever get out of a search tree.

So let's suppose I want to deal with a specific search tree; say I'm only searching for some given set of keywords. In a frame of mind where that's the most important problem in the world, then I use all my ammunition on that problem, to find as much as I can about it and try to push the limits. In the case where all the searches are actually unsuccessful, so that they end up at a leaf of the tree, the problem is mathematically similar to what is called Huffman coding: To find the shortest binary code that matches the statistical frequencies of the alphabet that you're encoding. But there's an extra twist: Your binary code must also preserve alphabetic order. This generalization of Huffman coding was studied by Gilbert and Moore in the 1950s.

On the other hand, maybe you have a case where all the searches are actually successful, yet some keys are much more likely than others. Iverson considered that problem [41, p.142–144]. And in my paper of 1971, I went further and considered the most general case, where some searches are successful and others are unsuccessful; we're given the probabilities of ending up at any particular node or at any particular leaf. What is the best search tree in that case? I found a way to construct it, in $O(n^2)$ steps.

Focusing on specific problems such as this makes you understand a larger territory of related structures and similar patterns. After I wrote this paper on binary search trees, I had become quite familiar with search trees; I had confidence that they weren't going to hold any surprises for me in the future.

Edgar: Continuing this topic of distinguishing between different use cases, in the following excerpt you distinguished between a first case in which traversal dominates and a second case in which insert and delete dominate:

> So a threaded binary tree is decidedly superior to an unthreaded one, with respect to traversal. These advantages are offset in some applications by the slightly increased time needed to insert and delete nodes in a threaded tree. [75, p.326]

Don: Good design is always a question of tradeoffs. If you're trying to minimize the time, do you want to put those threads in or not? The answer depends on the context of how you're using the tree. If certain operations are going to be common then you want the threads there; otherwise you don't. If insertions and deletions dominate instead of traversal then you wouldn't want to bother putting threads in. The context matters, and that's why I don't like using closed libraries. I like the ability to modify code.

Edgar: Suppose you've written your code and suppose the traversals dominate for one set of inputs but the insertions and deletions dominate for another set of inputs.

Don: I think it would be too hard to have a hybrid scheme that would switch over. You just have to make a choice and live with it.

Edgar: In your 1974 Turing Award lecture you said:

> [W]e have actually succeeded in making our discipline
> a science, and in a remarkably simple way: Merely by
> deciding to call it "computer science." [58, p.667]

Don: That was a joke, tongue in cheek.

Edgar: Your remark nicely summarizes the history of computing:
Some people actually did just call their discipline "computer science."

Don: In fact, a similar question came up last night as well: Why
can we call it computer science, when computers aren't
part of nature? I said, well, the name of something isn't
really important, the important thing is what computer
scientists do. That phrase, "computer science," became
popular almost instantly in America, because it rolls
naturally off a person's tongue in our language. Nobody
remembers what the word "mathematics" used to mean in
ancient times: We know what mathematics is, but we don't
know what the word meant in ancient Greek, nobody
remembers that.

The people who invented the words "computer science"
did, however, choose their words carefully. They correctly
said that science means knowledge about some area,
and that the rich phenomena surrounding computers are
definitely worthy of study. There's nothing about the word
"science" that confines it to the study of nature or about any
other particular kind of knowledge. And as I said earlier,
I've personally come to the conclusion that science is what
we understand well enough to explain to a machine.

Edgar: In your writings you mention that computer science
has a different meaning depending on who you talk
to [59, p.323]. This is again in line with a pluralistic
view on science and technology. I guess your definition
of computer science differs from, say, Stephen Cook's
definition.

Don: I see what you are saying.

In my Turing Award lecture [58] I made a detailed study of what the word "science" means as opposed to "art." That was part of my general way of explaining the world in which these words "art" and "science" live, and then to present a computer scientist's view of what it means and how it can be inspiring.

Edgar: You've repeatedly sought a balance between theory and practice, between the infinite and the finite. Or, as Dave Walden describes you on the official site of ACM Turing Award winners, you are "the rare theoretician who writes many lines of code every day."[1] In 1976 you wrote a paper entitled: 'Mathematics and Computer Science: Coping with Finiteness' [63].

Don: I wrote most of that paper in Montréal while I was giving the Aisenstadt lectures, late in 1975. The idea here is to get an understanding of what it means to be finite. People have used the term without knowing how big finite numbers can be. They say that if you limit yourself to finite things, you're just dealing with trivialities. In this paper I exhibit finite numbers that are so big that they cannot possibly be comprehended. They are finite, yet massively nontrivial. And almost all finite numbers are even bigger yet.

Edgar: This was still before you started working on TeX?

Don: I had no idea of TeX in 1975 and 1976. I was busily gathering material for Volume 4.

Edgar: In this 1976 paper on finiteness you also wrote about the Traveling Salesrep Problem:

> In fact, it may well be possible in a few years to prove that no good algorithm exists for the traveling salesman problem. Since so many people have tried for so many

---

[1] See *amturing.acm.org/award_winners/knuth_1013846.cfm*

> years to find a good algorithm, without success, the trend is now to look for a proof that success in this endeavor is impossible. [63, p.1240–1241]

> It is reasonably likely that, some day, somebody will prove that no good algorithm exists for this [traveling salesrep] problem. [63, p.1242]

In other words, in 1976 you still thought that P is not equal to NP.

Don: I was trying to present, to a general scientific audience, an example of something where we don't think we're going to have the breakthroughs that we've had in other areas. My opinion on this has flipped in an unusual way.

Edgar: But you flipped your opinion only after your work on TEX, right?

Don: Yes. I was working on TEX from 1977 onwards; meanwhile Garey, Johnson, and I were writing letters back and forth. [Don searches through the Garey & Johnson book [30].] The index of their book contains references to "private communications" with me. These would be letters where we'd be talking about problem SS4 on page 237: Sequencing to Minimize Weighted Completion Time. Similarly there was problem ND40 on page 217: Disjoint Connecting Paths. I'm pointing this out so that you can get a better feeling for those days; I didn't wait until this book came out before I knew these guys. We were all in correspondence with each other. Garey and Johnson's book was a masterpiece, it brought together the work of hundreds of people. And after it came out I actually wrote a joint paper with them, together also with Ron Graham [31]. It's a paper about algorithms to find the bandwidth of a graph.

Edgar: That's an NP-complete problem?

Don: Yes, but certain cases are solvable in polynomial time.

Edgar: You were looking at combinatorial problems, graph problems, and the like. [Don: Yes.] I want to show you some problems in the Garey & Johnson book which are about Loop programs, bit optimizations, and so on. [Edgar turns to page 275.]

Don: Right. The book covers algebra problems, geometry problems, and many others; this section is called 'Programs and Schemes.' Like when you're writing a compiler, questions about optimization occur and those can be NP-complete problems. Here's one where you want to check if programs are equivalent or not. (I'm not sure exactly why a compiler writer would want to do that, but sometimes you have to check the program structure and answer questions about it.) Ianov schemes on page 276 is a particular formalism for programming used in Russia in the early days. In other words, NP-hard problems arise in many different branches of knowledge.

Edgar: During our first interview we talked about program manipulation systems [88, p.60]. In your Turing Award lecture you also wrote:

> One thing all programmers have in common is that they enjoy working with machines; so let's keep them in the loop. [...] I have been trying to avoid misdirected automation for many years. [58, p.672]

Don: I was calling for interactive program transformation systems. My first major paper was about minimizing latency time when accessing data on a rotating drum [47]. I found that the best solution was not to automate the entire process but to have people in the loop. If a compiler optimizes a program, it is only optimizing one of many possible ways that the program could have been written down. But when you write the program, you know that many different programs would actually do the same thing, and you have no good way to tell the machine that those other programs would also be equally good for your purposes. The different programs aren't necessarily equivalent to each other in a strict sense, but they do give equivalent results on the data that you are concerned about. Therefore if people are involved in the loop, if people are helping to guide the optimization, then they can provide this extra knowledge.

Edgar: These results on page 275 in Garey and Johnson's book suggest that the task of building a completely automated program transformation system is an NP-complete problem at best. Several issues have been shown to be undecidable (e.g., [97, 111]).

Don: People have mistakenly believed that they should shy away from special cases of every hard problem. Now, at last, they're finally learning that such fears were unfounded. There are many examples now, with respect to satisfiability and other supposedly intractable problems, where industrial applications with more than a million variables are routinely solved, because those cases don't happen to be the hardest instances of the NP-hard problem whose reputed hardness used to scare people off. On the other hand, there also are instances with fewer than a hundred variables whose satisfiability or unsatisfiability is still unknown.

8. TEX & Literate Programming

Edgar: There are several interviews with you about TEX.[1] In hindsight, you were transferring the problem of analog typography to digital typography; that is, capturing everything mathematically.

Don: There was nothing profound about it. In 1976 I was getting galley proofs that looked awful, and at the beginning of 1977 I learned that changes in technology had converted the task of printing into a problem amenable to computer science. New devices allowed machines to place ink wherever you wanted to put it, on a grid. So the problem of printing had become a problem of constructing matrices of 0s and 1s, with 0s for blank spaces and 1s for ink. With good enough resolution (a large enough number of dots per centimeter), the results were every bit as good as could be obtained with traditional analog devices for printing.

I thought it would be pretty simple to take the letters in Thomas's calculus book — well, actually, in the first printing of Volume 1 of TAOCP — and to find out how wide the letters were in different places. I only had to capture that data in a way that could be expressed mathematically instead of as a photograph.

My first attempts were to use the TV camera at Stanford's AI Lab. A TV camera outputs screen images with shades of

---

[1]See Dave Walden's list at *tug.org/interviews/#knuth*

gray in a grid, and we had some primitive ways to capture those images. But that approach turned out to be terrible. For one thing, the lens of a TV camera doesn't take squares into squares. Even worse were the severe distortions of ambient light: If somebody illuminated the room just a little bit less or a little bit more, the letters would get 10% bigger or smaller. So it was impossible to get consistent results; some letters would come out looking much darker than others. And I didn't even know what the correct darkness was, because it depended on the glasses that I was wearing. So the whole idea of defining letterforms reliably via TV images was hopeless.

Better equipment was available at the laboratories of Xerox PARC, adjacent to the Stanford campus. I could go there and use their advanced video cameras; but then they said they would own everything I did. I would have had to assign the rights to everything that I acquired with their equipment, all the measurements would belong to them. That was unacceptable to me, because I wanted the letterforms to be expressed in a purely mathematical way that could be used freely by the world.

Then I tried to copy the letterforms myself, by hand, using simple equipment at home to magnify 35mm slides and project them on the wall. Jill had taken some good photos of the real pages that were used to print my book. So if I enlarged them, I thought I'd be able to trace the outlines, and I'd be able to adjust those outlines by tuning them afterward. But that approach failed, because the enlarged images were much too fuzzy.

However, while I was doing this and projecting on the wall and drawing the outlines, I started to put myself in the picture of the person who actually had drawn those letters for the first time when working for the Monotype Corporation. That person had sat down with an idea in mind: He had made the letter 'n' by making a stem and an arch, and he knew that he wanted to place the stems a certain distance apart. He had a definite system in his mind. So I realized that I wasn't just dealing with an image, I was dealing with something that was drawn with a plan.

My new goal, therefore, was not just to capture the image but to discover the plan. Comparing the letter 'n' to the letter 'h', I saw immediately that they're actually almost the same: They have the same kind of strokes, and their stems have the same relative positions. Similarly, the letter 'm' is just twice as long as the letter 'n'; well, the actual ratio is three to two, because the 'm' has three bars and 'n' has two. And the 'i' has one. So there was this logic in the original design that I could recapture.

I continued to dig further and to explore the underlying idea that lay behind these letters. I wound up getting access to Monotype's brass patterns, which were carved in metal and used actually with a machine to trace around the edge and make the molds that had created the actual type used in Thomas's calculus book.

Edgar: Changing a letter to boldface or italics made you realize that the underlying idea was not very systematic, right?

Don: Such changes in weight or slope are mostly linear and fairly easy to handle, but other issues crop up as well: For example, the serifs don't slope the same way. I wound up with about 65 different parameters that needed to be tuned, in order to make appropriate adjustments. I could change the parameters, and if that changed the letter 'a,' it would change the letter 'b' in a corresponding way. My challenge was to psych out how the designers had made these variations when they had produced type that was more or less bold, slanted or unslanted, larger or smaller.

In other words, my original idea was to express letters mathematically by measuring photographic images. That didn't work well, so I changed the whole idea to one of capturing the intelligence of the design instead of the outcome of the design. It's analogous to what we spoke of earlier, wanting to see the proof of a theorem instead of just the theorem. Like Pólya said: Find the methods and not the results. My quest for type design wasn't much different from my ordinary research activities.

Edgar: In retrospect, you mathematicized the whole problem but at a cost of using extremely many parameters. You also looked up the work of specialists, including Goudy and Zapf.

Don: Yes, Goudy and Zapf had written detailed books about their methods [35, 116]. Now, Mr. Goudy was very rigid and uncompromising; he basically didn't care for anybody else's work. I'm sure that he would have hated computers getting involved with it in any way whatsoever. Type design had to be done a certain way and that was his way. Hermann Zapf, on the other hand, was always open to innovations and further understanding of the processes. I got introduced to Hermann and then we worked together quite well. He had the most beautiful shapes.

In fact, I later dedicated *The METAFONTbook* [68] to him, and the byline is "to Hermann Zapf, whose strokes are the best." This byline has a double meaning: The first meaning is that he draws the most beautiful strokes. But the second meaning is more of an American colloquialism. When somebody gives you a compliment they call it a stroke, or a pat on the back. Hermann never would give a compliment unless he truly believed it. He would never say that something was good just to be nice. He would immediately tell you whenever he disliked anything. So he was the perfect critic, because I could trust that when he would approve of something I knew that I had finally gotten there. He would give me bad strokes when I had it wrong; but then when he finally would be happy with something, then I knew that I'd made some progress. So that's the double meaning in my dedication to him.

Edgar: During those ten years of TEX work, you also introduced a new programming methodology, called Literate Programming [64].

Don: Literate programming was born because of Tony Hoare's suggestion to publish the program for T_EX. Near the end of the 1970s I was telling people that I was working on typesetting, and I thought I could finish that project in a year. Tony Hoare was an editor for Prentice–Hall, and he was interested in putting out some books that would publish software. I don't think anything ever came from that project. He wanted to publish examples of good programming practice in the real world; such a thing had never yet been done, because it's too scary. I mean, most programs of industrial scale are so ugly that nobody wants to take responsibility for them and claim them as best practice. We had to meet real-world constraints, we were compromising all the time, so it was hard to put one's name on a program and say "this is the way everybody should do it." If you looked at essentially any program that was used in industry, you'd find that it sort-of worked, but you wouldn't really want to subject it to much scrutiny. Tony suggested though that I might think of publishing the program for T_EX, as part of a series of books that he was planning. And that was a very scary proposition. I thought it was probably impossible. On the other hand, I realized that it was a serious question. Why shouldn't it be possible to write large-scale programs that people could read?

I had come across a publication from Belgium, from Pierre de Marneffe, who had written a report called 'Holon Programming' [96]. I got hold of it somehow and decided to read it. I was bored by the first 99 pages, which seemed to be mostly empty philosophy. But then I got to page 100, where he started to give examples of what he called holon programming. And I realized that this would be an ideal way to present the program for T_EX. He demonstrated how to present a complicated program as a series of simple things related in simple ways. The reason I didn't like the first 99 pages was because they didn't have many examples in them; he understood the concepts at an abstract level, and he was right, but I didn't have any idea what he was talking about until I saw the concrete examples that came later.

So my life changed again, when I learned from de Marneffe the possibilities of breaking a program down into modules, where those modules can include both formal and informal descriptions. Each part of a "literate program" consists of a small piece of code preceded by a small piece of English. It has a fairly free form. The code might say that other code should be inserted here, and it might say that this code should be inserted elsewhere. So you can create a complex whole by creating a large number of small modules, each of which he called holons, each of which is related in a simple way to others. When everything is put together you have a piece of software that not only works but you can understand why it works, because all of its components are in digestible packages. That framework also scales up nicely from small projects to big ones.

Edgar: You already referred to de Marneffe's work in your 1974 paper 'Structured Programming with go to Statements' [61].

Don: Really? I would be very surprised if I had seen his work at that time. I thought his report came out later ...[2]

Peter Naur, of course, had been working on his snapshots and things like that. Peter had an early paper about programming by snapshots [99], bottom-up. Dijkstra's program for prime numbers and his top-down methods of structured programming were more recent [22]. Holons went to the next step, where I could see that I didn't have to be entirely bottom-up or top-down. Also, I could use TEX as an engine for making the final document. So everything came together and I began to write programs in a new

---

[2]Note by Knuth after returning home: By golly, you're right again! But in 1974 I didn't mention the holon concept; I cited his other remarks about integrating and optimizing different aspects of a design. I must have looked at his paper more than once.

way. I had played earlier with the methods of Dijkstra and Hoare [16], but the concept of literate programming gave those methods a new flavor.

Edgar: In the 1970s you were talking to Garey, Johnson, Karp, and others. Then, from 1977 till 1986, you worked on TEX. Was your TEX work really taking all of your energy?

Don: No, no. I was teaching classes, although not so often as before. I did write a dozen or so papers about typographical topics, but at least one purely algorithmic paper [66] was spawned as I completed the second edition of Volume 2 in 1981. There were others about verification of network protocols, about Huffman codes and balanced codes, an exposition of the theory of permanents, an analysis of cache memory, studies of continued fractions and random numbers, and even a paper about toilet paper [65]! In fact I also wrote about computer science history, an homage to the early IBM 650 computer on which I had cut my teeth [67].

During two or three of those years it's true that TEX was soaking up nearly all of my energy, but I was also keeping up with most other activities and teaching concrete math. I did teach a full-fledged class about METAFONT, as well as another one about the internals of TEX.

Edgar: Your work on TEX also led you to write your 1985 paper 'Optimal Prepaging and Font Caching' [29]. It's about communicating letter-shape information from a high-speed computer with a large memory to a type-setting device with a small memory. You made a distinction between a theoretical model and the actual program. In your words:

> [T]he theoretical model studied earlier [...] is a rather drastic simplification of the actual problem [...]. As usual. But (as usual) the theoretical considerations provided valuable guidelines for a practical implementation; and by using an algorithm that is optimal, or

> near-optimal under the simplifying assumptions, the authors were able to achieve quite satisfactory results, even though those assumptions were violated. [29, p.72–73]

Don: That's true. We often must simplify in order to gain high-level understanding. But the main contribution here was the prepaging. The classical algorithms that dealt with caching talked about page replacement. An operating system would realize that some data needed to be present in high-speed memory. If that memory was already filled to capacity it would have to decide which part of the old data was going to be the victim that goes away. The classical theorem said that if you knew the future, the victim should be the one that's not going to be needed for the longest period of time. Without such foreknowledge we might still have some heuristic information, so that we can make a decent guess about which ones are more or less likely; then we can devise a page replacement policy that works well.

In this paper Dave Fuchs and I added something to that well-known result. Instead of waiting until we have a page fault, at which time new data is desperately needed, suppose we try to anticipate the future and bring something into memory early on. If we're going to need ten new things in memory, we had better start inputting one of them now; otherwise they won't all be ready for us by the time we need them. It's called prepaging: getting ready for the future. We extended the known results about page replacement to say what's the best way to do prepaging, if you happen to know the future. And fortunately, our typesetting system actually did allow us to know the future, because it was a little "slave" algorithm residing in a tiny 8-bit computer, acting on instructions prepared by a "master" program on a mainframe that knew exactly what letters needed to be typeset and when. The high speed computer could easily look a page ahead, and see what letters were going to be needed on the next page. Thus it could start replacing the letters in the small computer's memory, as soon as those letters were no longer needed. It turned out that the same ideas became useful in operating systems later, when parallel machines became common.

9. P = NP

Edgar: One of the reasons why I just brought up your 1985 paper 'Optimal Prepaging and Font Caching' [29] is because you made a connection with the P versus NP question. In your words:

> Indeed, it would almost surely be unfeasible to develop an optimum strategy that takes account of all the details of the actual application, since the problem of optimum dynamic storage allocation is already NP-complete before we add the extra complexities of cache management (see [Garey & Johnson, 1979], problem SR2). [29, p.73]

Here you still thought that P is not equal to NP.

Don: Also today I don't disagree with this remark at all. Even though I now believe that P probably equals NP, I believe that there are two classes of problems: Those where we know of an explicit polynomial solution, and those where we know only that a polynomial solution exists. If it exists but we don't know what it is, it's no use to us. Therefore, instead of distinguishing between things that are in P and things that are not in P, I think we should distinguish between problems for which we know an algorithm and problems for which we don't. Cook and Karp's reducibility theory remains relevant, because we can say: If we would *know* a solution to the Traveling Salesrep Problem (TSP) then we would *know* a solution to the Satisfiability Problem (SAT). This is different from saying: If *there is* a polynomial time algorithm for TSP, then *there is* one for SAT. Existence of a polynomial time algorithm is not telling us that we could actually use it.

Edgar: That's like constructivism. You want a solution that includes the algorithm.

Don: Right. Just the fact that a polynomial time algorithm exists is not enough. Maybe the running time of that algorithm is still longer than the time since the beginning of the universe.

Edgar: But if you want to show that no polynomial time algorithm can solve SAT, then Cook's theory is suitable?

Don: Cook's theory is correct; it just needs to be correctly understood. If P is not NP, then of course no such algorithm exists. The point is that such a polynomial time algorithm could exist, and I think it probably does exist, because there are so many algorithms out there. Probably at least one of them works. But I suspect that we'll never know what those algorithms actually are.

Edgar: Were you already thinking along these lines when you wrote your 1985 paper?

Don: No, I wasn't thinking about that until much later.

Edgar: From your paper I conclude that you shared the general view on complexity theory in 1985.

Don: That's right. I was sharing the general view and I was still not aware of this important distinction between existence and knowledge of an algorithm.

Edgar: Isn't there a third possibility? Somebody could show that the P versus NP problem isn't so relevant after all. In other words, the problem will remain unresolved (for some time) but it won't remain an obstacle. The P versus NP problem is receiving a lot of attention now, but, once again, it's all based on the Turing machine model of computation.

Don: The Turing model is no different from almost any other model of sequential computation, with respect to the P versus NP question. Earlier today, when we were discussing program transformations, I alluded to the fact that people now are finding daily that the P versus NP distinction is irrelevant in many practical problems. A new generation of algorithms for satisfiability is able to resolve many nontrivial problems that we couldn't handle ten years ago. Those problems are special cases of an NP-hard problem, but we will likely never know in advance whether a particular special case is doable or not.

Many of these industrial problems are coming from people who have used their own expertise to design a computer system. They submit their design to a verifier, asking "Have I messed up? Is there going to be a race condition in this circuit or not?" The verifier (a SAT solver) has to check millions of cases, but it can report "No, there isn't a race condition." It's a nontrivial satisfiability problem and they get the answer.

A few weeks ago I played with satisfiability problems of a different kind. The question in this case was, "Is there a Boolean circuit that efficiently evaluates this function?" I had three functions of five variables for which I believed that the best way to evaluate them simultaneously would require 12 gates. But my program found a way to do it with 11 gates. I was extremely surprised, because I had believed for a long time that the existing 12-gate solution was so clever [85, p.108]. The fact that 13 gates weren't necessary was already quite amazing; but wow, we don't even need 12. The SAT solver came up with an 11-gate solution after four hours of computation; I don't think a human being would have been able to come up with it. It represents another breakthrough discovered by a machine.

So, we've got these examples of NP-hard problems where we have sophisticated algorithms that can see their way through all these cases. And it seems hard to believe that anybody will really understand why some of the problems are harder than others. We can try to keep learning more about them, but the holy grail of knowing which problems in P are actually nice and which problems in NP are actually nice — I expect that such methodology can be an art but not a science.

Edgar: Saying this means that the P versus NP problem remains relevant.

Don: It's relevant because it teaches us about the nature of computation. But it doesn't mean that we hold our breath until somebody has solved it; in that sense it not terribly relevant. If we were under attack by Martians, I don't think I would work on that particular problem.

Edgar: Peter Wegner has published articles in the *Communications of the ACM* and elsewhere in which he scrutinizes the Turing machine model of computation [34, 114, 115]. He proposes another, "more realistic" and "interactive" model of computation. An implication of his work is that the P versus NP problem could become much less relevant [23, Section 6] in another sense than yours. You already knew Peter Wegner in the 1960s. Could you tell me more about him? I believe Maurice Wilkes was his mentor.

Don: We became friendly in the 60s, and he had a major influence on my early work with attribute grammars (see [71]). I don't know of any Wilkes connection. Peter was educated at the London School of Economics; his early book [113] had a good deal of material about lambda expressions, which were definitely not a favorite topic for Maurice.

There was a tragic accident in 1999: Peter was hit by a bus while visiting London, and he lay unconscious in a coma for a long time. After many months of rehabilitation, guided by his faithful wife Judith who has also been a long-term friend, he has begun again to write papers. The main topic of those papers concerns the distinction between what I call an "algorithm" versus a "computational method" in TAOCP [75, p.5–8]. The former starts and stops; the latter can be ongoing, interactive, and reactive. I haven't looked more closely into the details of his recent publications.

Edgar: After your work on TEX you wrote in 1986 that you were "now turning back to [your] main life's work" which was writing about the "theoretical underpinnings of efficient computer programs" [45]. How do the following two writings fit into your main life's work:

- 'Nested Satisfiability' [70] from 1990?

- 'The Sandwich Theorem' [73] from 1994?

Don: The first one, 'Nested Satisfiability,' is a minor work that I thought was cute but not of great importance. I was really surprised to find that some other people have latched on to the topic and generalized the results of that paper. I thought of it as an academic exercise.

The second one was written mostly in Sweden while I was visiting Institut Mittag-Leffler at the end of 1991. I had just learned some striking new results of importance for Volume 4 of TAOCP, so I tried to organize the concepts and prepare an exposition of the subject. It relates to two fundamental properties of a graph, the "coloring number" and the "clique number," both of which are NP-hard to compute; yet there's another number, the "Lovász number," that lies between them and can be computed in polynomial time. The latter computations use semidefinite programming, a topic that was new to me. The paper didn't appear until 1994 because it wasn't really a piece of research, it was a tutorial. The *Electronic Journal of Combinatorics* was launched in 1994, and it provided an excellent outlet for such a tutorial.

As I continue collecting materials for TAOCP, I largely focus on problems that can be solved with polynomial algorithms that are not only known, their running time is known to be n^d for rather small d. On the other hand, I'm also writing about NP-hard problems like satisfiability, because they have mysteriously turned out to be solvable in many cases of great interest.

Edgar: I see that you had changed your position on the P versus NP problem by 1996, because in that year you stated the following in an interview:

> I have a feeling that someone might resolve the problem in the worst possible way, which is the following. Somebody will prove that P is equal to NP because there are only finitely many obstructions to it *not* being equal to NP. [Knuth's emphasis]
>
> The result would be that there is some polynomial such that we could solve all these problems in polynomial time. However, we won't know what the polynomial is, we just know that it exists.
>
> So maybe this will be n to the trillionth or something like that — but it'll be a polynomial. But we'll never be able to figure it out because it would probably take too long to find out what the polynomial is. But it does exist. Which means that the whole question P = NP was the wrong question! It might go that way.
>
> You see, even if you have something that takes 2^n steps and you compare it to something that takes n^{100}, then at least you can solve the 2^n one for n up to 20 or 30. But the n^{100} you can't even do for $n = 2$. So the degree of that polynomial is very important. There are so many algorithms out there, the task of showing that no polynomial ones exist is going to be very hard. [109, p.12]

Don: 1996, that's earlier than I thought. I should have referred to Michael Paterson, because in the back of my mind I was remembering that he had long ago raised the possibility that 'P = NP' might be the wrong question [98, p.184–185], although he had a different context in mind. Mike is a leading computer scientist, a top theoretician. He worked a lot with Michael Fischer in the early days, and later became head of the department at Warwick.

My goodness: I used the same example last night in my talk. Here I am 18 years later using an identical answer to a question! I say that there might be only finitely many obstructions to having P unequal to NP.

Edgar: Can you explain what you mean by that?

Don: Well, no, you see, I can't prove it. But, in my 'Twenty Questions' from 2014 [110] I try to explain the situation in more detail:

> I've come to believe that P = NP, namely that there does exist an integer M and an algorithm that will solve every n-bit problem belonging to the class NP in n^M elementary steps.
>
> Some of my reasoning is admittedly naïve: It's hard to believe that P ≠ NP and that so many brilliant people have failed to discover why. On the other hand if you imagine a number M that's finite but incredibly large—like say the number [. . .] discussed in my paper on "coping with finiteness"—then there's a humongous number of possible algorithms that do n^M bitwise or addition or shift operations on n given bits, and it's really hard to believe that all of those algorithms fail.
>
> My main point, however, is that I don't believe that the equality P = NP will turn out to be helpful even if it is proved, because such a proof will almost surely be nonconstructive. Although I think M probably exists, I also think human beings will never know such a value. I even suspect that nobody will even know an upper bound on M.

[...] Knowledge of the mere existence of an algorithm is completely different from the knowledge of an actual algorithm.

Edgar: In the rest of your 'Twenty Questions,' you refer to a 1995 result of Robertson and Seymour [102] that explicitly refers to a finite number of obstructions:

> Robertson and Seymour have proved a famous theorem in graph theory: Any class C of graphs that is closed under taking minors has a finite number of minor-minimal graphs. [...] Therefore there exists a polynomial-time algorithm to decide whether or not a given graph belongs to C: The algorithm checks that G doesn't contain any of C's minor-minimal graphs as a minor.

> But we don't know what that algorithm is, except for a few special classes, because the set of minor-minimal graphs is often unknown. The algorithm exists, but it's not known to be discoverable in finite time.

> This consequence of Robertson and Seymour's theorem definitely surprised me, when I learned about it while reading a paper by Lovász. And it tipped the balance, in my mind, toward the hypothesis that $P = NP$.

> The moral is that people should distinguish between known (or knowable) polynomial-time algorithms and arbitrary polynomial-time algorithms. People might never be able to implement a polynomial-time-worst-case algorithm for satisfiability, even though P happens to equal NP. [110]

Don: Robertson and Seymour showed in 1995 that there is some finite number of graphs, maybe ten trillion graphs, that would give us an algorithm to solve a certain classification problem if we only had a list of those graphs. But we don't know what those graphs are, nor even whether ten trillion is enough. Their result gives no bound on how big the list is. The theorem of Robertson and Seymour assures us that the list is finite, and that's all anybody knows. We could systematically go through all possible graphs, and some of them would be minor-minimal, but we'd never know when to stop. So here's a concrete case where we don't know the algorithm but we know that it exists.

Edgar: It's decidable.

Don: It's not only decidable, it's in P, because for any minor-minimal graph you can decide in polynomial time whether or not it is a minor of your graph. So we already know that some algorithms in P have no known polynomial-time implementation; we know this because of a proof that a certain list of excluded counterexamples is finite. That's sort of what I meant when I referred to "only finitely many obstructions to it not being equal to NP."

Then I proceeded to observe, somewhat obliquely, that a problem of size n can be represented as a string of n bits. Now consider all possible algorithms that take a string of bits and do a polynomial number of operations on bitstrings that each take polynomial time. (You might concatenate strings, multiply strings, divide strings, combine them with bitwise AND or OR or XOR, etc.) You're only doing a polynomial number of operations. If n is the length of the string, then maybe you'll allow yourself to do $n^{10^{10^{10^{10}}}}$ operations on those bits. At the end, you might end up with, say, an even or odd bit pattern, and it happens that it's even when there's no solution and odd when the answer is yes.

Okay, so all these possible algorithms are out there, each involving a deterministic, polynomial number of arbitrary bit operations. That's lots and lots of algorithms, and it's very likely that at least one of them is correct, essentially by chance, because the total number of bits that we started out with is pretty small. The bits only have to represent the Traveling Salesrep Problem (or another NP-complete problem). It's pretty unlikely that all of those algorithms fail. But, if one of them works, it's there, it exists, yet we'll never see it. That's the chasm between existence and knowledge.

Edgar: We'll never see the algorithm unless somebody like Ramanujan —

Don: — is able to run through all of these possibilities.

Edgar: If I'm not mistaken, Ramanujan didn't run through all possibilities, he just saw things.

Don: Well Ramanujan wasn't always right. . . .

Edgar: If there is one polynomial-time algorithm out there that works, then there must be infinitely many such algorithms. Shouldn't it then be "easy" to find such an algorithm if P equals NP?

Don: No, no! Infinity divided by infinity can be arbitrarily small. An infinite haystack can contain infinitely many needles, but that fact doesn't make the search for a needle any easier.

Edgar: To recapitulate, you changed your position on the P versus NP problem around 1996?

Don: Yes. In 1996 I had realized that existence is different from knowledge, but I didn't have any concrete way to back it up. Then Robertson and Seymour did their graph work. That work gave me a handle with which to exhibit a case where we do actually know the existence of polynomial-time algorithms that will probably never be humanly knowable.

Edgar: Wrapping up for today, you've also told me that you don't write research papers anymore.

Don: That's right. My first paper, written before going to college, was in MAD magazine, which is a humor magazine. After more than 50 years, MAD is still going strong, and it's published in other countries too now. I got paid 25 dollars for that article; the full story is in Chapter 1 of my book *Selected Papers on Fun and Games* [84].

My final paper was another attempt at humor, which I called 'An Earthshaking Announcement' [3]. It's the last chapter of that book. It's about an ultimate version of TEX that not only does typesetting but also does everything else under the sun. It includes 3D printing, manufacturing, distribution, formula manipulation, social networking, videos, etc., and it also has a scheme for everybody to benefit without paying a nickel. It is going to be the answer to everybody's prayers, and do everything that every other system does, but do it better.

In that paper I pointed out that I've finally abandoned the foolish ways by which I used to do things: Instead of writing any of the programs myself, I've delegated the whole thing to a group of developers in the former Soviet Union, where I've secretly started up a small company. That group is writing the new system in Scheme, and using all of the buzzword-compliant techniques of software engineering that have ever been invented — except that the ideas of literate programming have been abandoned as unworkable.

That's the story of my successor to TEX. I presented it in San Francisco, near the place where Steve Jobs would introduce his new machines for Apple. I called it an earthshaking announcement, because earthquakes tend to happen in San Francisco. I also explained that I had planned to present a live demo, but unfortunately the implementors ran into a glitch and the new system wasn't quite ready yet. I spoke in that vein for 30 minutes, and the transcript of that talk is my last paper. My publishing career in papers is complete.

Edgar: Really?

Don: Yes. I'm not going to write any more papers, but I'm still writing books.

Edgar: Maybe if you stumble upon a proof of the P versus NP problem you'll want to publish that in a paper.

Don: No, I don't think so. I would just blog about it instead.

Bibliography

[1] A.V. Aho, J.E. Hopcroft, and J.D. Ullman. *The Design and Analysis of Computer Algorithms*. Ed. by M.A. Harrison. Addison-Wesley, 1974.

[2] D.J. Albers and G.L. Alexanderson, eds. *Mathematical People: Profiles and Interviews*. Boston: Birkhauser, 1985.

[3] "An Earthshaking Announcement". In: *TUGboat*. TUGboat 31.2 (2010), pp. 121–124. Reprinted as Chapter 49 of [84].

[4] J.-A.M. Anderson, L.M. Berc, J. Dean, S. Ghemawat, M.R. Henzinger, S.-T. Leung, R.L. Sites, M.T. Vandevoorde, C.A. Waldspurger, and W.E. Weihl. "Continuous Profiling: Where Have All the Cycles Gone?" In: *ACM Transactions on Computer Systems* 15.4 (1997), pp. 357–390.

[5] I. Asimov. *The Intelligent Man's Guide to Science. Volume 1: Physical sciences. Volume 2: Biological sciences*. New York: Basic Books, 1960.

[6] R. Braden. "Burroughs Algol at Stanford University, 1960–1963". In: *IEEE Annals of the History of Computing* 35.4 (2013), pp. 69–73.

[7] N.G. de Bruijn. *Asymptotic Methods in Analysis*. Amsterdam: North-Holland, 1958.

[8] A.W. Burks and H. Wang. "The Logic of Automata — Part I". In: *Journal of the ACM* 4.2 (1957), pp. 193–218.

[9] N. Chomsky. *Syntactic Structures*. The Hague/Paris: Mouton, 1957.

[10] A. Cobham. "The intrinsic computational complexity of functions". In: *1964 International Congress on Logic,*

Methodology and Philosophy of Science. Amsterdam: North-Holland, 1965, pp. 24–30.

[11] M.E. Conway. "Design of a separable transition-diagram compiler". In: *Communications of the ACM* 6.7 (1963), pp. 396–408.

[12] S.A. Cook. "Variations on Pushdown Machines (Detailed Abstract)". In: *Proceedings of the 1st Annual ACM Symposium on Theory of Computing, May 5–7, 1969, Marina del Rey, CA, USA.* 1969, pp. 229–231.

[13] S.A. Cook. "The Complexity of Theorem-Proving Procedures". In: *Proc. 3rd Annual ACM Symposium on Theory of Computing.* Association for Computing Machinery. 1971, pp. 151–158.

[14] S.A. Cook. "Linear time simulation of deterministic two-way pushdown automata". In: *Information Processing 71, Proceedings of IFIP Congress 71.* Amsterdam: North-Holland, 1972, pp. 75–80.

[15] "CSTUG, Charles University, Prague, March 1996: Questions and Answers with Prof. Donald E. Knuth". In: *TUGboat.* TUGboat 17.4 (1996), pp. 355–367. Reprinted with enhancements as Chapter 32 of [77].

[16] O.-J. Dahl, E.W. Dijkstra, and C.A.R. Hoare. *Structured Programming.* London/New York: Academic Press, 1972.

[17] O.-J. Dahl and K. Nygaard. "SIMULA—an ALGOL-based simulation language". In: *Communications of the ACM* 9.9 (1966), pp. 671–678.

[18] E.G. Daylight. *Pluralism in Software Engineering: Turing Award Winner Peter Naur Explains.* Ed. by E.G. Daylight and P. Naur; www.lonelyscholar.com; ISBN 9789491386008. Heverlee: Lonely Scholar, Oct. 2011.

[19] E.G. Daylight. *The Dawn of Software Engineering: from Turing to Dijkstra.* Ed. by K. De Grave; www.lonelyscholar.com; ISBN 9789491386022. Heverlee: Lonely Scholar, 2012.

[20] H.B. Demuth. "Electronic Data Sorting". PhD thesis. Stanford University, 1956.

[21] E.W. Dijkstra. "A Note on Two Problems in Connexion with Graphs". In: *Numerische Mathematik* 1 (1959), pp. 269–271.

[22] E.W. Dijkstra. *Notes on Structured Programming*. Tech. rep. T.H.-Report 70-WSK-03. Second edition. Published as Chapter 1 of [16]. Technische Hogeschool Eindhoven, Apr. 1970.

[23] E. Eberbach, D. Goldin, and P. Wegner. "Alan Turing: Life and Legacy of a Great Thinker". In: ed. by C. Teuscher. Springer, 2004. Chap. Turing's ideas and models of computation.

[24] J. Edmonds. "Paths, trees, and flowers". In: *Canadian Journal of Mathematics* 17 (1965), pp. 449–467.

[25] R.W. Floyd. "A descriptive language for symbol manipulation". In: *Journal of the ACM* 8.4 (1961), pp. 579–584.

[26] R.W. Floyd. "Bounded context syntactic analysis". In: *Communications of the ACM* 7.2 (1964), pp. 62–67.

[27] R.W. Floyd. "The syntax of programming languages—A survey". In: *IEEE Transactions on Electronic Computers* EC-13.4 (1964), pp. 346–353.

[28] E.H. Friend. "Sorting on Electronic Computer Systems". In: *Journal of the ACM* 3.3 (1956), pp. 134–168. ISSN: 0004-5411.

[29] D.R. Fuchs and D.E. Knuth. "Optimal Prepaging and Font Caching". In: *ACM Transactions on Programming Languages and Systems* 7.1 (1985), pp. 62–79. Reprinted and revised with an addendum as Chapter 14 of [82].

[30] M.R. Garey and D.S. Johnson. *Computers and Intractability: A Guide to the Theory of NP-Completeness*. W.H. Freeman and Company, 1979.

[31] M.R. Garey, R.L. Graham, D.S. Johnson, and D.E. Knuth. "Complexity results for bandwidth minimization". In: *SIAM Journal on Applied Mathematics* 34 (1978), pp. 477–495.

[32] J. von zur Gathen. "Who Was Who in Polynomial Factorization: 1". In: *Proceedings of the 2006 International Symposium on Symbolic and Algebraic Computation*. ISSAC '06. 2006, p. 2. ISBN: 1-59593-276-3.

[33] J. von zur Gathen and J. Gerhard. *Modern Computer Algebra*. Cambridge: Cambridge University Press, 1999.

[34] D.Q. Goldin and P. Wegner. "The Interactive Nature of Computing: Refuting the Strong Church-Turing Thesis". In: *Minds and Machines* 18.1 (2008), pp. 17–38.

[35] F.W. Goudy. *Typologia: Studies in Type Design & Type Making with Comments on the Invention of Typography, the First Types, Legibility & Fine Printing*. Berkeley, California: University of California Press, 1940.

[36] R. Graham, D.E. Knuth, and O. Patashnik. *Concrete Mathematics*. 1st ed. Reading, Massachusetts: Addison-Wesley, 1989.

[37] R. Graham, D.E. Knuth, and O. Patashnik. *Concrete Mathematics: A Foundation for Computer Science*. 2nd ed. Reading, Massachusetts: Addison-Wesley, 1994.

[38] J.M. Hammersley. "On the enfeeblement of mathematical skills by "Modern Mathematics" and by similar soft intellectual trash in schools and universities". In: *Bulletin of the Institute for Mathematics and Its Applications* 4 (1968), pp. 66–85.

[39] J. Hartmanis and R.E. Stearns. "On the computational complexity of algorithms". In: *Transactions of the American Mathematical Society* 117 (1965), pp. 285–306.

[40] C.A.R. Hoare. "Quicksort". In: *The Computer Journal* 5.1 (1962), pp. 10–16.

[41] K.E. Iverson. *A Programming Language*. John Wiley and Sons, Inc., 1962.

[42] G.B. Thomas Jr. *Calculus and Analytic Geometry: Functions of One Variable and Analytic Geometry*. Third edition, second printing. Reading, Massachusetts: Addison-Wesley, 1961.

[43] R.M. Karp. "Reducibility Among Combinatorial Problems". In: *Complexity of Computer Computations*. Ed. by R.E. Miller and J.W. Thatcher. London/New York: Plenum Press, 1972, pp. 85–103.

[44] M. Kauers and P. Paule. *The Concrete Tetrahedron: Symbolic Sums, Recurrence Equations, Generating Functions, Asymptotic*

Estimates. Text and Monographs in Symbolic Computation. Wien: Springer-Verlag, 2011.

[45]   D.E. Knuth. *Theory and Practice*. An after-dinner talk given at the Franklin Institute on 16 October 1986, in connection with a special convocation marking the fortieth anniversary of the creation of ENIAC. The transcript of the talk appears in Chapter 8 of [74].

[46]   D.E. Knuth. "Editorial (about technical writing)". In: *Engineering and Science Review* 2.3 (1959), p. 5. Case Institute of Technology. Reprinted as Chapter 4 of [83].

[47]   D.E. Knuth. "Minimizing drum latency time". In: *Journal of the ACM* 8 (1961), pp. 119–150. Reprinted with an addendum as Chapter 28 of [82].

[48]   D.E. Knuth. In: *Computing Reviews* 3.4 (1962). Review number 2140, p. 148.

[49]   D.E. Knuth. "A History of Writing Compilers". In: *Computers and Automation* 11.12 (1962), pp. 8–18. Reprinted with corrections and an addendum as Chapter 20 of [80].

[50]   D.E. Knuth. "Finite Semifields and Projective Planes". PhD thesis. Caltech, 1963.

[51]   D.E. Knuth. "An Empirical Study of FORTRAN Programs". In: *Software, Practice and Experience* 1.2 (1971), pp. 105–133. Reprinted with corrections and an addendum as Chapter 24 of [80].

[52]   D.E. Knuth. "Optimum Binary Search Trees". In: *Acta Informatica* 1 (1971), pp. 14–25. Reprinted with an addendum as Chapter 4 of [82].

[53]   D.E. Knuth. "The Dangers of Computer-Science Theory". In: *Logic, Methodology and Philosophy of Science IV*. Studies in Logic and the Foundations of Mathematics 74. Amsterdam: North-Holland, 1971, pp. 189–195. Reprinted with corrections and an addendum as Chapter 2 of [78].

[54]   D.E. Knuth. "Ancient Babylonian Algorithms". In: *Communications of the ACM* 15.7 (1972), pp. 671–677. Reprinted with corrections and an addendum as Chapter 11 of [74].

[55] D.E. Knuth. "George Forsythe and the Development of Computer Science". In: *Communications of the ACM* 15.8 (1972), pp. 721–726. Reprinted as Chapter 16 of [74].

[56] D.E. Knuth. "Mathematical analysis of algorithms". In: *Information Processing 71: Proceedings of IFIP Congress 1971, Vol 1*. Amsterdam: North-Holland, 1972, pp. 19–27. Reprinted as Chapter 1 of [78].

[57] D.E. Knuth. "A Terminological Proposal". In: *SIGACT News* 6.1 (1974), pp. 12–18. Reprinted as Chapter 28 of [78].

[58] D.E. Knuth. "Computer Programming as an Art". In: *Communications of the ACM* 17.12 (1974), pp. 667–673. Reprinted as Chapter 1 of [72].

[59] D.E. Knuth. "Computer Science and its Relation to Mathematics". In: *The American Mathematical Monthly* 81 (1974), pp. 323–343. Reprinted with an additional reference as Chapter 1 of [74].

[60] D.E. Knuth. "Postscript About NP-Hard Problems". In: *SIGACT News* 6.2 (1974), pp. 15–16. Reprinted as Chapter 29 of [78].

[61] D.E. Knuth. "Structured Programming with go to Statements". In: *Computing Surveys* 6.4 (Dec. 1974), pp. 261–301. Reprinted with corrections and an addendum as Chapter 2 of [72].

[62] D.E. Knuth. *Surreal Numbers*. Reading, Massachusetts: Addison-Wesley, 1974.

[63] D.E. Knuth. "Mathematics and Computer Science: Coping with Finiteness". In: *Science* 194 (1976), pp. 1235–1242. Reprinted with corrections and an addendum as Chapter 2 of [74].

[64] D.E. Knuth. "Literate Programming". In: *The Computer Journal* 27 (1984), pp. 97–111. Reprinted as Chapter 4 of [72].

[65] D.E. Knuth. "The toilet paper problem". In: *American Mathematical Monthly* 91.8 (1984), pp. 465–470. Reprinted with additional material as Chapter 16 of [78].

[66] D.E. Knuth. "Deciphering a linear congruential encryption". In: *IEEE Transactions on Information Theory* 31.1 (1985), pp. 49–52. Reprinted with an addendum as Chapter 24 of [82].

[67] D.E. Knuth. "The IBM 650: An Appreciation from the Field". In: *IEEE Annals of the History of Computing* 8.1 (1986), pp. 50–55. Reprinted with corrections as Chapter 13 of [74].

[68] D.E. Knuth. *The METAFONTbook*. Reading, Massachusetts: Addison-Wesley, 1986.

[69] D.E. Knuth. "Algorithmic Themes". In: *A Century of Mathematics in America*. Ed. by P.L. Duren. Vol. 1. Providence, Rhode Island: American Mathematical Society, 1988, pp. 439–445. Written for the 100th birthday of the American Mathematical Society. Reprinted as Chapter 5 in [74].

[70] D.E. Knuth. "Nested Satisfiability". In: *Acta Informatica* 28.1 (1990), pp. 1–6. Reprinted with corrections and an addendum as Chapter 8 of [82].

[71] D.E. Knuth. "The Genesis of Attribute Grammars". In: *Lecture Notes in Computer Science* 461 (1990), pp. 1–12. Keynote address at the International Conference on Attribute Grammars and their Applications, Paris, France, September 1990. Reprinted as Chapter 19 of [80].

[72] D.E. Knuth. *Literate Programming*. Vol. 27. CSLI Lecture Notes. Stanford, California: CSLI Publications, 1992.

[73] D.E. Knuth. "The Sandwich Theorem". In: *Electronic Journal of Combinatorics* 1 (1994). Reprinted as Chapter 8 of [81].

[74] D.E. Knuth. *Selected Papers on Computer Science*. Vol. 59. CSLI Lecture Notes. Stanford, California: CSLI Publications, 1996.

[75] D.E. Knuth. *The Art of Computer Programming, Volume 1: Fundamental Algorithms*. 3rd ed. Reading, Massachusetts: Addison-Wesley, 1997. ISBN: 0-201-89683-4.

[76] D.E. Knuth. "Teach Calculus with Big O". In: *Notices of the American Mathematical Society* 45.6 (1998), pp. 687–688. Published in abridged form, as a letter to the editor. Also printed as Chapter 3 of [83].

[77] D.E. Knuth. *Digital Typography*. Vol. 78. CSLI Lecture Notes. Stanford, California: CSLI Publications, 1999, xvi+685pp.

[78] D.E. Knuth. *Selected Papers on Analysis of Algorithms*. Vol. 102. CSLI Lecture Notes. Stanford, California: CSLI Publications, 2000.

[79] D.E. Knuth. "Robert W Floyd, In Memoriam". In: *SIGACT News* 34.4 (2003), pp. 3–13. Reprinted as Chapter 1 of [82].

[80] D.E. Knuth. *Selected Papers on Computer Languages*. Vol. 139. CSLI Lecture Notes. Stanford, California: CSLI Publications, 2003.

[81] D.E. Knuth. *Selected Papers on Discrete Mathematics*. Vol. 106. CSLI Lecture Notes. Stanford, California: CSLI Lecture Notes, 2003, xvi+812pp.

[82] D.E. Knuth. *Selected Papers on Design of Algorithms*. Vol. 191. CSLI Lecture Notes. Stanford, California: CSLI Publications, 2010.

[83] D.E. Knuth. *Companion to the Papers of Donald Knuth*. Vol. 202. CSLI Lecture Notes. Stanford, California: CSLI Publications, 2011.

[84] D.E. Knuth. *Selected Papers on Fun and Games*. Vol. 192. CSLI Lecture Notes. Stanford, California: CSLI Publications, 2011.

[85] D.E. Knuth. *The Art of Computer Programming, Volume 4A: Combinatorial Algorithms, Part I*. Upper Saddle River, New Jersey: Addison-Wesley, 2011.

[86] D.E. Knuth. *Let's Not Dumb Down the History of Computer Science*. Presentation by Knuth on Wednesday 7 May 2014 in the Mackenzie Room, Huang Building 300, 4:00 pm, Stanford University. kailathlecture.stanford.edu/2014KailathLecture.html. 2014.

[87] D.E. Knuth and R.H. Bigelow. "Programming Languages for Automata". In: *Journal of the ACM* 14.4 (1967), pp. 615–635. Reprinted with corrections and an addendum as Chapter 12 of [80].

[88] D.E. Knuth and E.G. Daylight. *The Essential Knuth*. Ed. by K. De Grave. Conversations. Heverlee: Lonely Scholar, 2013. ISBN: 9789491386039.

[89] D.E. Knuth and S. Gorn. "Backus' Language". In: *Communications of the ACM* 5.4 (1962), p. 185. ISSN: 0001-0782. DOI: 10.1145/366920.366925.

[90] D.E. Knuth, J.H. Morris Jr., and V.R. Pratt. "Fast Pattern Matching in Strings". In: *SIAM Journal on Computing* 6 (1977), pp. 323–350. Reprinted with corrections and an addendum as Chapter 9 of [82].

[91] D.E. Knuth and L. Trabb Pardo. "The Early Development of Programming Languages". In: *Encyclopedia of Computer Science and Technology*. Ed. by J. Belzer, A.G. Holzman, and A. Kent. Vol. 7. New York: Marcel Dekker, 1977, pp. 419–493. Reprinted with corrections and an addendum as Chapter 1 of [80].

[92] D.E. Knuth and F. Ruskey. "From Object-Orientation to Formal Methods: Essays in Memory of Ole-Johan Dahl". In: ed. by O. Owe, S. Krogdahl, and T. Lyche. Vol. 2635. Lecture Notes in Computer Science. Heidelberg: Springer-Verlag, 2004. Chap. Efficient coroutine generation of constrained Gray sequences, pp. 183–204. Reprinted as Chapter 25 of [80].

[93] W.C. Lynch. "Ambiguities in Backus Normal Form languages". PhD thesis. Madison, Wisconsin: University of Wisconsin–Madison, 1963.

[94] M.S. Mahoney. *Histories of Computing*. Ed. by T. Haigh. Cambridge, Massachusetts/London, England: Harvard University Press, 2011.

[95] A.A. Markov. *Theory of Algorithms*. Vol. 42. Trudy Matematicheskogo Instituta imeni V. A. Steklova. Moscow/Leningrad: Academy of Sciences of the USSR, 1954.

[96] P.-A. de Marneffe. *Holon programming: A survey*. Tech. rep. 135 pp. Université de Liège, Service Informatique, 1973.

[97] A.R. Meyer and D.M. Ritchie. "The complexity of loop programs". In: *Proceedings of the ACM National Meeting*. 1967, pp. 465–469.

[98] R.E. Miller (moderator), C.M. Fiduccia, R. Floyd, J.E. Hopcroft, R.M. Karp, M. Paterson, M.O. Rabin, V. Strassen, and S. Winograd. "Panel discussion (transcript)". In: *Complexity of Computer Computations*. Ed. by R.E. Miller and J.W. Thatcher. London/New York: Plenum Press, 1972, pp. 169–185.

[99] P. Naur. "Proof of Algorithms by General Snapshots". In: *BIT Nordisk Tidskrift for Informationsbehandling* 6 (1966), pp. 310–316. Reprinted in Section 5.2 of [100].

[100] P. Naur. *Computing: A Human Activity*. New York: ACM Press/Addison-Wesley, 1992.

[101] M.O. Rabin and D. Scott. "Finite Automata and their Decision Problems". In: *IBM Journal of Research and Development* 3.2 (1959), pp. 114–125.

[102] N. Robertson and P. Seymour. "Graph Minors. XIII. The disjoint paths problem". In: *Journal of Combinatorial Theory, Series B* 61.1 (1995), pp. 65–110.

[103] B. Salvy, B. Sedgewick, M. Soria, W. Szpankowski, and B. Vallée. "Obituary. Philippe Flajolet". In: *Journal of Symbolic Computation* 46.9 (2011), pp. 1085–1086.

[104] J. Shallit. "Origins of the analysis of the Euclidean algorithm". In: *Historia Mathematica* 21.4 (1994), pp. 401–419.

[105] W.G. Spohn Jr. "Can mathematics be saved?" In: *Notices of the American Mathematical Society* 16.6 (1969), pp. 890–894.

[106] W. Szpankowski. *Analysis of Algorithms (AofA): Part I: 1993 – 1998 ("Dagstuhl Period")*. Unpublished, available from: citeseerx.ist.psu.edu/viewdoc/summary?doi= 10.1.1.8.3985.

[107] R.E. Tarjan. "Efficiency of a Good But Not Linear Set Union Algorithm". In: *Journal of the ACM* 22.2 (1975), pp. 215–225. ISSN: 0004-5411.

[108] R.E. Tarjan. "A Class of Algorithms which Require Nonlinear Time to Maintain Disjoint Sets". In: *Journal of Computer and System Sciences* 18.2 (1979), pp. 110–127.

[109] "TUG'95: Questions and Answers with Prof. Donald E. Knuth". In: *TUGboat*. TUGboat 17.1 (1996). Reprinted as Chapter 31 of [77], pp. 7–22.

[110] *Twenty Questions for Donald Knuth, 20 May 2014* (2014). www.informit.com/articles/article.aspx?p=2213858.

[111] J. Voeten. "On the fundamental limitations of transformational design". In: *ACM Transactions on Design Automation of Electronic Systems* 6.4 (2001), pp. 533–552.

[112] H. Wang. "Games, Logic and Computers". In: *Scientific American* 213.5 (1965), pp. 98–106.

[113] P. Wegner. *Programming Languages, Information Structures, and Machine Organization*. New York: McGraw-Hill, 1968.

[114] P. Wegner and D.Q. Goldin. "Computation beyond Turing machines". In: *Communications of the ACM* 46.4 (2003), pp. 100–102.

[115] P. Wegner and D.Q. Goldin. "Principles of problem solving". In: *Communications of the ACM* 49.7 (2006), pp. 27–29.

[116] H. Zapf. *About Alphabets: Some Marginal Notes on Type Design*. Cambridge, Massachusetts: M.I.T. Press, 1970.

Index

ACM, 1, 3, 19, 41, 43, 66, 76, 92

Addison–Wesley, 18, 19

Africa, 7

Aho, Alfred V., 1, 37, 38, 42, 44, 57

AIP, 15

algebra, 26, 78

ALGOL, 33, 66

ALGOL-W, 45, 46

Alpha, 69

America, 64, 75

American, 15, 20, 59, 64, 83

analysis of algorithms, 1, 4, 5, 7, 21, 27, 41, 44, 59–61

anthropomorphic, 47

Apple, 99

Armenia, 7

Art of Computer Programming, The, 3, 5, 10, 17, 19, 20, 24, 25, 27, 28, 31, 32, 34, 43–45, 52, 59, 60, 69, 80, 93

artificial intelligence, 4, 33

Asimov, Isaac, 12, 13

asymptotic, 2, 24, 26, 27, 29, 35, 36, 69

    analysis, 21, 24, 27

    complexity, 23

    estimate, 24

    game, 23, 60

    method, 1, 4, 20, 21, 24, 28, 32, 36, 60

    running time, 23

Australia, 7

Austria, 62

AUTOMATH, 22

automaton, 20, 32, 38, 39, 48–50, 53, 69

    automata theorist, 50

    automata theory, 20, 42, 48–51, 69

    linking, 34, 35, 37

    pushdown, 49

Babylon, 55

Backus, John W., 43

BALGOL, 33, 34

Barcelona, 8

basket weaving, 13

Belgium, 6, 84

Bellman, Richard, 5

Berkeley, 7, 52

Berlin, 8

big-Oh, 24

Bigelow, Richard H., 48

Bourbaki, Nicolas, 1, 59–61

Brownian motion, 6

bubble sort, 26, 31

Bucharest, 70, 71

Burks, Arthur W., 20

Burroughs, 3, 33

California, 3, 33, 56, 57

Caltech (California Institute of Technology), 20, 22, 59, 64
Case Institute of Technology, 3, 15, 27, 33, 43
Caswell, Herb, 19
Ceauşescu, Nicolae, 71
Cheriton, David R., 30
China, 7
Chomsky, 118
Chomsky, Avram N., 43, 44
Cobham, Alan J., 32
combinatorial
    algorithm, 53, 56
    mathematics, 11
    method, 4
    pattern, 11
    problem, 32, 78
    seminar, 53
compiler, 5, 27, 28, 33, 34, 41, 42, 44–46, 78
complexity
    theorist, 29, 32, 39
    theory, 5, 29, 34, 42, 90
concrete, 59, 61, 85, 97, 98
    mathematics, 1, 25, 59–62, 86
    tetrahedron, 62
Conway, Mel, 43
Cook, Stephen, 1, 20, 32, 49–54, 57, 75, 89, 90
culture, 13, 14
Cummings, Lew Addison, 18
Cummings, Melbourne Wesley, 18
curriculum, 16
Cæsar, Julius, 4

Dagstuhl, 6

Dahl, Ole-Johan, 47, 65–67
de Bruijn, Nicolaas G., 20–22, 25, 27
Demuth, Howard B., 19, 42
Dijkstra, Edsger W., 17, 66, 67, 86
Dutch, 22, 66
dynamic programming, 5

economic, 15, 92
Edmonds, Jack R., 32, 56
Egypt, 7
Enflo, Per H., 58
Euclid, 16
Europe, 4, 6, 8

fable, 14
Feigenbaum, Edward A., 65
Feller, William, 28
Fibonacci numbers, 16
Fischer, Michael J., 95
Flajolet, Philippe, 4, 6, 28, 40
Floyd, Robert W, 17, 41, 44–47, 66
Forsythe, George E., 55
FORTRAN, 27, 33, 44, 68
Fox, Robert, 63
France, 60, 61
French, 6, 71
Friend, Edward H., 19
Frontiers of Knowledge Award, 3
Fuchs, David R., 87

Gödel, Kurt F., 67
Gardner, Martin, 64, 65
Garey, Michael R., 54, 55, 77–79, 86, 89
Gaul, 4

Gerhard, Jürgen, 16
Germany, 6
Google, 72
Gorn, Saul, 43, 44
Goudy, Frederick W., 12, 83
grammar, 42–45, 92
    context free, 41, 42, 44
Great Britain, 64
Greenwich, 9
Grimm, Jacob and Wilhelm, 14

Hartmanis, Juris, 1, 32, 34, 37
hash, 27, 28, 42, 55
hemp, 14
Hennie-Stearns construction, 69
history, 9, 10, 12–16, 33, 53, 57,
        61, 63, 64, 68, 70, 75, 86
    external, 13, 14, 63
    historian, 1, 9, 15, 16, 37, 64,
        65
    internal, 13, 14, 63
Hoare, Charles A.R., 29, 67, 84,
        86
Hopcroft, John E., 37, 38, 57
Huffman code, 73, 86
humanities, 14

IBM, 53
    IBM 650, 86
IFIP, 5, 65
Illustrated London Times, The,
        64
infinity, 23, 24, 27, 29, 34, 35, 39,
        98
Internet, 17
interpreter, 45
IPL, 33
Israel, 7
Iverson, Kenneth E., 26, 27, 73

James, William, 71, 72
Jobs, Steven P., 99
Johnson, David S., 54, 55, 77–79,
        86, 89

Karp, Richard M., 30, 51–55, 86,
        89
Kauers, Manuel, 4, 62
Klop, Jan Willem, 65
Knuth-Morris-Pratt algorithm,
        67
Kyoto Prize, 3

lg, 31, 35
lg lg, 23
linear probing, 27, 28
LISP, 33
little-oh, 24
log, 23, 29, 38, 55
log log, 24, 39
London, 8, 64, 92
Louchard, Guy, 6
Lovász
    László, 96
    number, 93
LR(k), 44, 45
Luxemburg, Wilhelmus, 20
Lynch, William C., 44

MAD, 98
Madison, Wisconsin, 44
Mahoney, Michael S., 2, 37, 42
Markov, 19
    algorithm, 57
    Andrei, 19
    finite state language, 43
    process, 19
Martian, 58, 92
Mazur, Stanisław, 58

Mellin transform, 22
METAFONT, 83, 86
Milwaukee, Wisconsin, 3
MIT (Massachusetts Institute of Technology), 18
Monotype Corporation, 19, 81, 82
Moore's law, 55
Morocco, 7
Morris, Jim H., 49, 51, 67
mosque, 6

Naur, Peter, 11, 67, 72, 85
Netherlands, The, 20, 22, 65
New York Times, The, 64
New Zealand, 7
Norway, 54–56, 58, 65, 66
Norwegian, 66
NP-complete, 55, 57, 77–79, 89, 97
NP-hard, 55–57, 78, 79, 91, 93, 94
numerical analysis, 4
Nygaard, Kristen, 66

object oriented, 22, 46, 66
operations research, 5
Oslo, 56, 57, 65
Oxford, 63

P=NP question, 1, 58, 89–92, 94, 95, 98, 100
Pólya, George, 11, 83
palindrome, 49, 50
palstar, 49, 50
Paris, 4, 6–8, 29, 60
parser, 41, 42, 44–46
Pasadena, 3
Pascal, 46

Paterson, Michael S., 94
pattern matching, 50, 67
Paule, Peter, 62
physicist, 15, 65
physics, 14, 15
pi, 14
pointer, 1, 34, 35, 37, 46, 55
    machine, 32, 34, 38, 39
Poland, 58
police, 8
polynomial, 94
    complete, 57
    factorization, 16
    time, 32, 39, 77, 89, 90, 93, 94, 96–98
polyomino tiling, 22
Pratt, Vaughan R., 7, 51, 52, 58, 60, 67
Prentice–Hall, 84
Princeton, 16, 28
publisher, 18, 19

queuing theory, 28, 42
quicksort, 5, 29, 31

Ramanujan, Srinivasa, 98
random access, 37–39, 50
recursive
    coroutine, 46, 47, 66
    descent, 45, 46
    formula, 21
    function, 37
Robertson, Neil, 96–98
Romania, 71
Royal Society, 63

San Francisco, 8, 99
Sarnak, Peter, 60
satisfiability, 32, 52–54, 58, 79, 89, 91, 93, 94, 96

Schützenberger, Marcel-Paul, 71
Scheme, 99
science, 118
Sedgewick, Robert, 60
Seymour, Paul D., 96–98
Shallit, Jeffrey O., 16
SIGCIS, 9
SIMULA, 47, 66
sorting, 19, 29, 31, 42
South America, 7
Southeast Asia, 6
Soviet Union, 99
stack, 29, 46, 48
Stanford University, 1, 3, 7, 9, 11, 16, 29, 30, 34, 52, 58–61, 65, 80, 81
Stearns, Richard E., 32, 34, 37
Stirling number, 61
summation, 4, 62
Sweden, 58, 93

Tarjan, Robert E., 1, 23, 28–30, 37–39, 55
taxi, 4, 8
TEX, 1, 12, 76, 77, 80, 83–86, 93, 99
Thomas, George B., Jr., 18, 19, 80, 82
Tjome, 56
Toronto, 52
Traveling Salesrep Problem, 53, 76, 77, 89, 97
trie, 20
Turing, 49
    Award, 1, 3, 68, 75, 76, 78
    machine, 2, 34–37, 39, 49, 57, 58, 69, 90, 92
    model, 91
typesetting, 2, 19, 84, 87, 99

Uber, 8
Ullman, Jeffrey D., 37, 38, 57
Univac SS90, 44
universe, 23, 24, 90
Uzbekistan, 7

von zur Gathen, Joachim, 16

Walden, David, 2, 76, 80
Wang, Hao, 20
Warwick, 95
Washington D.C., 8
Wegner, Peter, 92
Weiner, Charles, 15, 65
Wilkes, Maurice V., 92
Wirth, Niklaus E., 45, 46, 67
Wisconsin, 3
Wolf Composition Company, 19
Wolf, Hans, 19
World War II, 21

Zapf, Hermann, 12, 83

Also by Edgar G. Daylight:

Conversations

- Pluralism in Software Engineering:
  Turing Award Winner Peter Naur Explains
  2011 · ISBN 9789491386008

- Panel discussions I & II, held at the Future of Software
  Engineering Symposium
  2011 · ISBN 9789491386015

- The Essential Knuth
  2013 · ISBN 9789491386039

Full-length books

- The Dawn of Software Engineering:
  from Turing to Dijkstra
  2012 · ISBN 9789491386022

Find our latest publications at www.lonelyscholar.com.

LONELY SCHOLAR™
SCIENTIFIC BOOKS

From the same authors:

The Essential Knuth

Abstract

Donald E. Knuth lived two separate lives in the late 1950s. During daylight he ran down the visible and respectable lane of mathematics. During nighttime, he trod the unpaved road of computer programming and compiler writing.

Both roads intersected! — as Knuth discovered while reading Noam Chomsky's book *Syntactic Structures* on his honeymoon in 1961.

> Chomsky's theories fascinated me, because they were mathematical yet they could also be understood with my programmer's intuition. It was very curious because otherwise, as a mathematician, I was doing integrals or maybe was learning about Fermat's number theory, but I wasn't manipulating symbols the way I did when I was writing a compiler. With Chomsky, wow, I was actually doing mathematics and computer science simultaneously.

How, when, and why did mathematics and computing converge for Knuth? To what extent did logic and Turing machines appear on his radar screen?

The early years of convergence ended with the advent of Structured Programming in the late 1960s. How did that affect his later work on TeX? And what did "structure" come to mean to Knuth?

Shedding light on where computer science stands today by investigating Knuth's past — that's what this booklet is about.